P9-DVN-894

Complex Networks

Structure, Robustness and Function

Examining important results and analytical techniques, this graduate-level textbook is a step-by-step presentation of the structure and function of complex networks.

From the stability of the Internet to efficient methods of immunizing populations, from epidemic spreading to how to efficiently search for individuals, this textbook explains the theoretical methods used, and the experimental and analytical results obtained. Ideal for graduate students and researchers entering this field, it gives detailed derivations of many results in complex networks theory. End-of-chapter review questions help students monitor their understanding of the materials presented.

Reuven Cohen is a Senior Lecturer in the Department of Mathematics at Bar-Ilan University, Israel. He has written many papers in the fields of complex networks, robot swarms, algorithms and communication networks, and has won several national and international prizes for his work.

Shlomo Havlin is a Professor in the Department of Physics at Bar-Ilan University, Israel. He is an Editor of several physics journals, has published over 600 articles in international journals, co-authored and co-edited 11 books, and won numerous awards for his work including the Weizmann Prize (2009) and the APS Lilienfeld Prize (2010).

Complex Networks

Structure, Robustness and Function

REUVEN COHEN
Bar-Ilan University

SHLOMO HAVLIN
Bar-Ilan University

CAMBRIDGE UNIVERSITY PRESS
Cambridge, New York, Melbourne, Madrid, Cape Town, Singapore,
São Paulo, Delhi, Dubai, Tokyo

Cambridge University Press
The Edinburgh Building, Cambridge CB2 8RU, UK

Published in the United States of America by Cambridge University Press, New York

www.cambridge.org
Information on this title: www.cambridge.org/9780521841566

First published 2010

Printed in the United Kingdom at the University Press, Cambridge

A catalog record for this publication is available from the British Library

Library of Congress Cataloguing in Publication data
Cohen, Reuven, 1974–
Complex networks : structure, robustness, and function / Reuven Cohen, Shlomo Havlin.
p. cm.
Includes bibliographical references and index.
ISBN 978-0-521-84156-6 (hardback)
1. System analysis. I. Havlin, Shlomo. II. Title.
T57.6.C638 2010
003 – dc22 2010007255

ISBN 978-0-521-84156-6 Hardback

Contents

1 Introduction

Networks are present in almost every aspect of our life. The technological world surrounding us is full of networks. Communication networks consisting of telephones and cellular phones, the electrical power grid, computer communication networks, airline networks and, in particular, the world-wide Internet network are an important part of everyday life. The symbolic network of HTML pages and links – the World Wide Web (WWW) – is a virtual network that many of us use every day, and the list is long. Society is also networked. The network of friendship between individuals, working relations, or common hobbies, and the network of business relations between people and firms are examples of social and economic networks. Cities and countries are connected by road or airline networks. Epidemics spread in population networks. A great deal of interest has recently focused on biological networks representing the interactions between genes and proteins in our body. Ecological networks such as predator–prey networks are also under intensive study today. The physical world is also rich in network phenomena such as interactions between atoms in matter, between monomers in polymers, between grains in granular media, and the network of relations between similar configurations of proteins (i.e. between configurations that are in reach of each other by a simple move). Recently, studies have shown that polymer networks in real space can actually have a wide distribution of the branching factor, which is also similar to other real-world networks [ZKM^{+}03].

Graphs are used for describing mathematical concepts in networks. Graphs represent the essential topological properties of a network by treating the network as a collection of nodes and edges. For example, in computer networks, such as the Internet, computers can be represented by nodes, and the cables between them are represented by the edges. In the WWW the nodes are the HTML pages, and the edges represent the links between pages. This is a simple, yet powerful concept. Because of its simplicity it considers different complex systems, such as those described above, using the same mathematical tools and methods and, in many cases, the properties of the networks are similar.

Graph theory is rooted in the eighteenth century, beginning with the work of Euler, who is the father of the field of topology as well as many other fields in mathematics. The theory began with the famous problem of the bridges of Kőnigsberg, where people had been wondering for years whether all seven bridges connecting the different parts

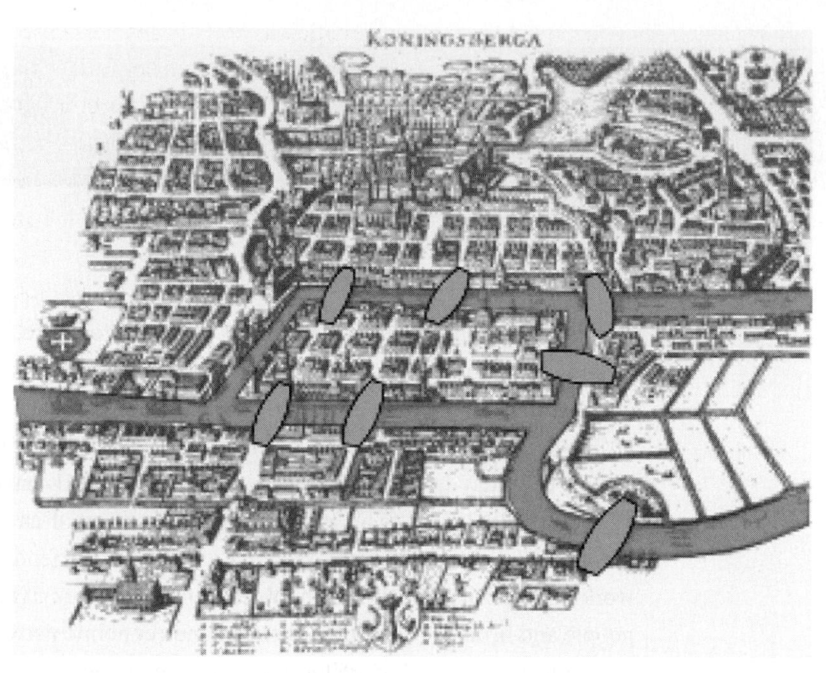

Figure 1.1 The bridges of Kőnigsberg (after Wikipedia).

of the town could be traversed, without passing any of them twice (see Figure 1.1). The genius of Euler led him to the understanding that the only important factor in this problem is the topological network structure, and therefore it can be simplified into a graph traversal problem, containing nodes (parts of the city) and links (bridges). He then proceeded to solve the problem by concluding that to fulfill the requirement every node in the graph, except possibly the first and last nodes visited, should be connected by an even number of bridges (since it is entered and left the same number of times). In Kőnigsberg more than two nodes have an odd number of links, and therefore the bridges cannot be traversed by such a path, known thereafter as an Eulerian path.

This simple yet powerful argument shows the strength of graph theory, enabling deduction of properties of real-world systems using simplification in order to construct a very basic model. Studies of graph theory usually focus on the properties of special graphs or on extremal properties (finding graphs with extreme properties). However, the networks mentioned above are hardly appropriate for such research. They change over time, social links are created and broken, technological networks are changed daily by the addition of new nodes, as are the links between them. Biological networks change by evolutionary processes and by environmental processes. Even at a given time point, one cannot usually find the complete data for the network structure.

In the 1960s, two mathematicians, Paul Erdős and Alfred Rényi (ER), introduced a new concept that allows the treatment of such networks – random graph theory [ER59, ER60].[1] Their ingenious idea was to combine the concepts of graph theory with tools from probability theory and to consider families of graphs rather than specific graphs. Random graph theory is to graph theory what statistical mechanics is to Newtonian physics. The microscopic theory underlies the small-scale behavior, but when the entire ensemble is considered, new statistical concepts and collective behavior emerge.

The study of random graphs has led to ideas very similar to those of statistical physics. Since statistical physics deals with a system of many interacting atoms and molecules it is natural to assume that methods from this field will be useful for network study. Indeed, percolation, scaling, order parameters, renormalization, self-similarity, phase transitions, and critical exponents from statistical physics are all present in the field of random graphs, and are used in studying such networks.

At the end of the twentieth century, with the advancement of computers and the availability of large-scale data and the tools to analyze them, it became clear that the classical theory of random graphs fails to describe many real-world networks. The works of Barabási and Albert on the WWW [BA99] and of Faloutsos, Faloutsos, and Faloutsos on the Internet router network [FFF99] made clear that the link distribution of these and many other networks is not completely random, and it cannot be described by ER graph theory. These findings and others have led to a new, generalized form of random graph theory, taking into account some less trivial correlations found in real-world networks. These results explain several long-standing puzzles, for example, why viruses and worms are able to survive in the Internet for a very long time. Moreover, studying these new types of networks leads to novel physical laws, which arise owing to the new topology. If materials such as polymers can be constructed with a similar topology, it is expected that they will obey new and anomalous physical laws such as phase transitions, elasticity, and transport.

This book will focus on this modern theory of complex networks. Since thousands of papers, as well as several popular science [Bar03, Wat03] and scientific books [BBV08, DM03, PV03] have appeared on this subject in the last few years, it will be impossible to cover all existing works. In this book we have tried to focus on results concerning the structure of these networks and also partially cover works regarding the dynamics and applications. Since this is also intended as a textbook for students and scientists aiming to enter this growing field, we will attempt to present a detailed and clear description of the methods used in analyzing complex networks. This, we believe, will allow the reader to obtain further results in this growing field and to comprehend further literature on this subject.

[1] In fact, some of these ideas had been raised before, in particular, by Rapoport [Rap57]. However, only with the systematic works of Erdős and Rényi was much attention given to this subject.

The rest of this introduction will present some basic definitions and concepts from physics and mathematics. The main body of this book is divided into three parts. Part I will present results based on measurements in real-world networks, and will present several ensembles and growth models studied in this field. Part II will discuss the structural and robustness properties of complex networks. It will focus mainly on scale-free networks, which are thought to be most relevant for real-world systems, but in most cases this approach is also suitable for other types of random networks. Part III will discuss some dynamics regarding complex networks, and applications of the knowledge gained to real-world problems. The appendices will provide more technical details regarding probability theory, as well as algorithmic and simulation aspects.

1.1 Graph theory

A **graph** according to its mathematical definition is a pair of sets (V, E), where V is a set of vertices (the nodes of the graph), and E is a set of edges, denoting the links between the vertices. Each edge consists of a pair of vertices and can be regarded as similar to "bonds" in physical systems.

In a **directed graph** (also termed "digraph"), the edges are taken as ordered pairs, i.e., each edge is directed from the first to the second vertex of the pair.

A "**multigraph**" is a graph in which more than one edge is allowed between a pair of vertices and edges are also allowed to connect a vertex to itself. This is less restrictive than the notion of a graph, and therefore many of the networks studied in this work will actually be multigraphs.

A graph is represented frequently by an **adjacency matrix**, A_{ij}, which is a matrix in which every row and column represents a vertex of the graph. The A_{ij} entry is 1 if a link exists between the ith and jth vertices, or 0 otherwise. In a directed graph, the matrix will, in general, be asymmetrical. In a multigraph the entries can also be integers larger than 1, and the diagonal entries are not necessarily 0.

1.2 Scale-free processes and fractal structures

In statistical physics, it is well known that systems approaching a critical point in a phase transition develop a behavior that spans all length-scales of the system. Close to criticality, the correlations between physically remote regimes change from decaying exponentially with the distance, to a slow, power-law, decaying behavior.

This power-law phenomenon has no characteristic length-scale, and is therefore often termed "scale free." The reaction to external disturbances, for example, the susceptibility of the system, also diverges as a power law when approaching the critical point. Another situation where power laws and scale-free behavior appear is in self-organized criticality (SOC) [BTW87], where events such as earthquakes and forest fires tend to drive themselves into a criticality-like power-law behavior.

Power-law distributions have been studied in physics, particularly in the context of fractals and Lévy flights. Fractals are objects having no characteristic length-scale and appear similar (at least in a statistical sense) at every length-scale [BBV08, BH94, BH96, bH00, BLW94, Fed88, Man82]. Many natural objects, such as mountains, clouds, coastlines and rivers, as well as the cardiovascular and nervous systems are known to be fractals and are self-similar. This is why we find it hard to distinguish between a photograph of a mountain and part of the mountain; neither can we ascertain the altitude from which a picture of a coastline was taken. Diverse phenomena, such as the distribution of earthquakes, biological rhythms, and rates of transport of data packets in communication networks, are also known to possess a power-law distribution. They come in all sizes and rhythms, spanning many orders of magnitude [BH96].

Lévy flights were suggested by Paul Lévy [Lév25], who was studying what is now known as Lévy stable distributions. The question he asked was, when is the length distribution of a single step in a random walk similar to that of the entire walk? Besides the known result, that of the Gaussian distribution, Lévy found an entire new family – essentially that of scale-free distributions. Stable distributions do not obey the central limit theorem (stating that the sum of a large number of steps, having *finite variance*, tends to a Gaussian distribution [Fel68]), owing to the divergence of the variance of individual steps. Lévy walks have numerous applications [GHB08, HBG06, KSZ96, SK85, SZK93]. An interesting observation is that animal foraging patterns that follow Lévy stable distributions have been shown to be the most efficient strategy [Kle00a, VBH$^+$99]. For recent reviews and books on complex networks and, in particular, scale-free networks, see [AB02, BBV08, BLM$^+$06, DG08, DM02, DMS03, New02b, PV03].

PART I

RANDOM NETWORK MODELS

2 The Erdős–Rényi models

Before 1960, graph theory mainly dealt with the properties of specific individual graphs. In the 1960s, Paul Erdős and Alfred Rényi initiated a systematic study of random graphs [ER59, ER60, ER61]. Some results regarding random graphs were reported even earlier by Rapoport [Rap57]. Random graph theory is, in fact, not the study of individual graphs, but the study of a statistical ensemble of graphs (or, as mathematicians prefer to call it, a *probability space* of graphs). The ensemble is a class consisting of many different graphs, where each graph has a probability attached to it. A property studied is said to exist with probability P if the total probability of a graph in the ensemble possessing that property is P (or the total fraction of graphs in the ensemble that has this property is P). This approach allows the use of probability theory in conjunction with discrete mathematics for studying graph ensembles. A property is said to exist for a class of graphs if the fraction of graphs in the ensemble which does not have this property is of zero measure. This is usually termed as a property of **almost every (a.e.)** graph. Sometimes the terms "almost surely" or "with high probability" are also used (with the former usually taken to mean that the residual probability vanishes exponentially with the system size).

2.1 Erdős–Rényi graphs

Two well-studied graph ensembles are $G_{N,M}$ – the ensemble of all graphs having N vertices and M edges, and $G_{N,p}$ – the ensemble consisting of graphs with N vertices, where each possible edge is realized with probability p. These two families, initially studied by Erdős and Rényi, are known to be similar if $M = \binom{N}{2}p$, so long as p is not too close to 0 or 1 [Bol85]; they are referred to as ER graphs. These families are quite similar to the microcanonical and canonical ensembles studied in statistical physics.

Examples of other well-studied ensembles are the family of **regular graphs**, where all nodes have the same number of edges, $P(k) = \delta_{k,k_0}$, and the family of *unlabeled* graphs, where graphs that are isomorphic under permutations of their nodes are considered to be the same object.

An important attribute of a graph is the average degree, i.e., the average number of edges connected to each node. We will denote the degree of the ith node by k_i and the average degree by $\langle k \rangle$. N-vertex graphs with $\langle k \rangle = \mathcal{O}(N^0)$ are called **sparse graphs**. In this book, we concern ourselves mainly with sparse graphs.

An interesting characteristic of the ensemble $G_{N,p}$ is that many of its properties have a related **threshold function**, $p_\mathrm{t}(N)$, such that the property exists, in the "thermodynamic limit" of $N \to \infty$, with probability 0 if $p < p_\mathrm{t}$, and with probability 1 if $p > p_\mathrm{t}$. This phenomenon is the same as the physical concept of a percolation *phase transition*. An example of such a property is the existence of a **giant component**, i.e., a set of connected nodes, in the sense that a path exists between any two of them, whose size is proportional to N. Erdős and Rényi showed [ER60] that for ER graphs, such a component exists if $\langle k \rangle > 1$. If $\langle k \rangle < 1$, only small components exist, and the size of the largest component is proportional to $\ln N$. Exactly at the threshold, $\langle k \rangle = 1$, a component whose size is proportional to $N^{2/3}$ emerges. This phenomenon was described by Erdős as the "double jump."[1] We will later relate the exponent $2/3$ to the fractal dimension of the incipient infinite cluster (giant component) at criticality, studied in the statistical physics literature [BH96, SA94]. It can also be shown that at criticality the size distribution of the components (clusters) is $n(s) \sim s^{-2.5}$ [BH96, SA94]. Another property is the average path length between any two nodes, which in almost every graph of the ensemble (with $\langle k \rangle > 1$ and finite) is of order $\ln N$. The small, logarithmic distance is actually the origin of the "small-world" phenomena that characterize networks.

2.2 Scale-free networks

The Erdős–Rényi model has traditionally been the dominant subject of study in the field of random graphs. Recently, however, several studies of real-world networks have found that the ER model fails to reproduce many of their observed properties.

One of the simplest properties of a network that can be measured directly is the degree distribution, or the fraction $P(k)$ of nodes having k connections (degree k). A well-known result for ER networks is that the degree distribution is Poissonian,

$$P(k) = \mathrm{e}^{-z} z^k / k!, \tag{2.1}$$

where $z = \langle k \rangle$ is the average degree [Bol85].

[1] For a discussion of the phase transition in random graphs see, e.g., [AS00, Bol85].

Direct measurements of the degree distribution for real networks, such as the Internet [AB02, FFF99], WWW (where hypertext links constitute directed edges) [BAJ00, BKM+00], email network [EMB02], citations of scientific articles [Red98], metabolic networks [JMBO01, JTA+00], trust networks [GGA+02], airline networks [BBPV04], neuronal networks [ECC+05], and many more, show that the Poisson law does not apply (for more details, see Chapter 3). Rather, often these nets exhibit a scale-free degree distribution:

$$P(k) = ck^{-\gamma}, \qquad k = m, \ldots, K \tag{2.2}$$

where $c \approx (\gamma - 1)m^{\gamma-1}$ is a normalization factor, and m and K are the lower and upper cutoffs for the degree of a node, respectively. The divergence of moments higher than $\lceil \gamma - 1 \rceil$ (as $K \to \infty$ when $N \to \infty$) is responsible for many of the anomalous properties attributed to scale-free networks.

All real-world networks are finite and therefore all their moments are finite. The actual value of the cutoff K plays an important role. It may be approximated by noting that the total probability of nodes with $k > K$ is of order $1/N$ [CEbH00, DMS01c]:

$$\int_K^\infty P(k)\,\mathrm{d}k \sim 1/N. \tag{2.3}$$

This yields the result

$$K \sim mN^{1/(\gamma-1)}. \tag{2.4}$$

The degree distribution alone is not enough to characterize the network. There are many other quantities, such as the degree-degree correlation (between connected nodes), the spatial correlations, the clustering coefficient, the betweenness or centrality distribution, and the self-similarity exponents. These quantities will be defined in Chapter 3.

Several models have been presented for the evolution of scale-free networks, each of which may lead to a different ensemble. The first proposal was the *preferential attachment* model of Barabási and Albert, which is known as the "Barabási–Albert model" [BA99]. Several variants of this model have been suggested (see, e.g., [BB01, KRL00]). In this book we will focus on the configuration model, also known as the "Bollobás construction" [Bol85], (sometimes also referred to as the "Molloy–Reed construction" [ACL00, MR95, MR98]), which ignores the evolution and assumes only the degree distribution and no correlations between nodes. Thus, the node reached by following a link is independent of the origin. For algorithms that generate networks with a given degree distribution, see Appendix C.

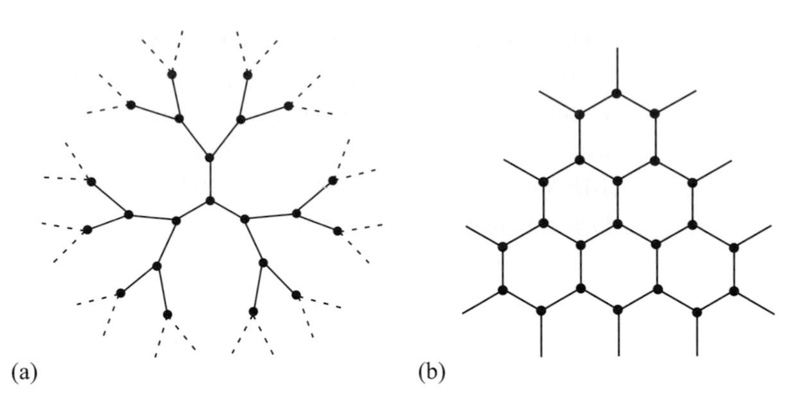

Figure 2.1 (a) A Cayley tree with $z = 3$. (b) A hexagonal lattice with the same degree.

2.3 Diameter and fractal dimensions

Regular lattices can be viewed as networks embedded in Euclidean space, of a well-defined dimension, d. This means that $n(r)$, the number of nodes within a distance r from an origin, grows as $n(r) \sim r^d$ (for large r). For fractal objects, d in the last relation may be a non-integer and is replaced by the fractal dimension d_f. Similarly, the chemical dimension, d_l, is defined by the scaling of the number of nodes within l edges (the shortest distance along edges) from a given node (an origin), $n(l) \sim l^{d_l}$. A third dimension, $d_{\min}$, relates between the chemical path length, l, and Euclidean distances, $l \sim r^{d_{\min}}$. It satisfies $d_{\min} = d_\mathrm{f}/d_l$ [BH94, BH96, bH00].

An example of a network where the above power laws are not valid is the Cayley tree (also known as the Bethe lattice). The Cayley tree is a regular graph, of fixed degree z, and no loops (see Figure 2.1(a)). It has been studied by physicists in many contexts, since its simplicity often allows for exact analysis. An infinite Cayley tree cannot be embedded in a Euclidean space of finite dimensionality. The number of nodes at l is $n(l) \sim (z-1)^l$. Since the exponential growth is faster than any power law, Cayley trees are referred to as *infinite-dimensional* systems.

In most random network models, the structure is locally tree-like (since most loops occur only for $n(l) \sim N$), and since the number of nodes grows as $n(l) \sim \langle k-1 \rangle^l$, they are also infinite dimensional. As a consequence, the diameter of such graphs (i.e., the minimal path between the most distant nodes) scales as $D \sim \ln N$ [Bol85]. Many properties of ER networks, including the logarithmic diameter, are also present in Cayley trees. This small diameter in ER graphs and Cayley trees is in contrast to that of finite-dimensional lattices, where $D \sim N^{1/d_l}$.

Similar to ER, percolation on infinite-dimensional lattices and the Cayley tree (studied as early as 1941 by Flory [Flo41]) yields a critical threshold $p_\mathrm{c} = 1/(z-1)$.

For $p > p_c$, a "giant cluster" of order N exists, whereas for $p < p_c$, only small clusters appear. For infinite-dimensional lattices[2] (similar to ER networks) at criticality, $p = p_c$, the giant component is of size $N^{2/3}$ [BH96]. This last result follows from the fact that percolation on lattices in dimension $d \geq d_c = 6$ is in the same universality class as infinite-dimensional percolation, where the fractal dimension of the giant cluster is $d_f = 4$, and therefore the size of the giant cluster scales as $N^{d_f/d_c} = N^{2/3}$. The dimension d_c is called the "upper critical dimension." Such an upper critical dimension exists not only in percolation phenomena, but also in other physical models, such as in the self-avoiding walk model for polymers and in the Ising model for magnetism; in both these cases $d_c = 4$.

Watts and Strogatz [Wat99, WS98] suggested a model that retains the local high clustering of lattices (i.e., the neighbors of a node have a much higher probability of being neighbors than in random graphs) while reducing the diameter to $D \sim \ln N$. This so-called, "small-world network" is achieved by replacing a fraction ϕ of the links in a regular lattice with random links, to random distant neighbors. (In other variants of the small-world model the "long-range" links are simply added on, without prior removal of lattice links, see Figure 4.1.) A study of scale-free networks embedded in Euclidean space (at the obvious price of an upper cutoff for k) that exhibit finite dimensions can be found in [bRCH03, MS02, RCbH02, WSS02b] and will be discussed in Section 4.6.

2.4 Random graphs as a model of real networks

Many natural and man-made systems are networks, i.e., they consist of objects and interactions between them. These include computer networks, in particular the Internet, logical networks, such as links between WWW pages, and email networks, where a link represents the presence of a person's address in another person's address book. Social interactions in populations, work relations, etc. can also be modeled by a network structure. Interactions between chemicals, genes, proteins, neurons, airports, companies, etc. can also be modeled this way. Networks can also describe possible actions or movements of a system in a configuration space (a phase space), and the nearest configurations are connected by a link.

[2] For Cayley trees some complications arise, since the structure of a tree presents some critical behavior even at $p = 1$, although $p_c = 1/(z-1)$. As will be understood from this book, the properties of random networks, and in particular random regular networks, are more appealing as models of infinite-dimensional systems and are more well behaved than the Cayley tree model. For example, the giant component in a Cayley tree at criticality does not scale as $N^{2/3}$, but is much smaller, of order of $\log^2 N$ due to the absence of loops.

All the above examples and many others have a graph structure that can be studied. Many of them have some ordered structure, derived from geographical or geometrical considerations, cluster and group formation, or other specific properties. Physical systems, such as solid materials, are modeled by lattice networks, owing to their structure. This is why in physics many approaches, such as percolation, have been developed for lattice models. However, most of the above networks are far from regular lattices and are much more complex and random in structure. Therefore, it is plausible that they maintain many properties of the appropriate random graph model. We will focus in this book on the topological properties of random graphs with a given degree distribution, and in particular scale-free graphs. We will also present several practical applications of functional properties.

2.5 Outlook and applications

We will show throughout this book that in many aspects scale-free networks can be regarded as a generalization of ER networks. For large γ (usually, for $\gamma > 4$) the properties of scale-free networks, such as distances, optimal paths, and percolation, are the same as in ER networks. In contrast, for $\gamma < 4$, these properties are very different and can be regarded as anomalous. The anomalous behavior of scale-free networks is due to the strong heterogeneity in the degree of the nodes, which breaks the node-to-node translational homogeneity (symmetry) that exists in the classical homogeneous networks, such as lattices, Cayley trees, and ER graphs. The small variation of the degrees in the ER model or in scale-free networks with large γ is insufficient to break this symmetry, and therefore many results for ER networks are the same as for Cayley trees, where the degree of each node is the same.

The knowledge gained in studying the topological and dynamical properties of complex networks, and in particular scale-free networks, can be applied to describe many properties of real networks in diverse fields.

In Chapter 6 we discuss the average distance between nodes in scale-free networks. We show that the distance $\gamma \leq 3$ is even smaller than in regular random graphs, and hence it is given the name "ultra small worlds." For $\gamma > 3$, the distances scale logarithmically with N as in ER networks, characteristic of small worlds, whereas for $\gamma < 3$ the distance scales as $\log \log N$. We also present studies of the layer structure (tomography) of scale-free networks, and suggest a construction of the lowest possible diameter for scale-free networks. This has implications for subjects such as searching in physical and logical networks [ALPH01], packet routing in the Internet, and transport of information [GKK01b], as well as the structure of multicast trees in the Internet [KCM$^+$06]. The knowledge gained will hopefully enable, for example, the analysis and design of better, more efficient algorithms for searching, routing, immunization, and traffic control in large-scale computer networks.

Real networks are dynamical in character. The most frequent change in computer network structure is the breakdown of nodes and links due to malfunction, maintenance, or intentional hacker attacks. In Chapter 10, we discuss the influence of such events on the network's structure and integrity. We present analytical and simulation results, and discuss general methods for studying the cluster sizes in the network. One surprising result is that random scale-free networks with $\gamma < 3$ are robust against random failure of any fraction $q < 1$ of the nodes. This is in contrast with ER networks and scale-free networks with $\gamma > 3$, where there is a finite threshold, $q_c < 1$, and when $q > q_c$ of the nodes is randomly removed, the network disintegrates. This robustness against random failure is an important advantage of scale-free networks over regular networks. We also present results for the behavior at and before the network's disintegration. These results may be applied to study efficient routing and traffic control in the presence of crashes. It may also be applied for designing more robust network structures, and it has a strong connection to epidemic and virus spreading in networks (discussed also in Chapter 15).

Many natural and man-made networks are, in fact, directed. Several examples are the WWW, email networks, metabolic networks (where a chemical may act as a catalyst for another one, but not vice versa), and economical networks (where trade relations are not symmetrical). In Chapter 11, we study the percolation properties of such networks. We discuss the importance of correlations between the *in*- and *out*-degrees and the rich structure of the critical behavior near the percolation threshold. This has implications regarding the stability and survivability of directed networks, on virus spreading through directed networks (such as email viruses), and on the stability of biochemical processes, which are fundamental to life.

The heterogeneous degree topology of real and logical computer networks and population networks has also been shown to lead to very low epidemic thresholds [AM92, PV01b] and to render random immunization strategies inefficient [AM92, CEbH00, PV02]. This understanding has led to the suggestion of using targeted immunization strategies of the highest degree nodes [DB02, PV02]. In Chapter 15, we present an alternative efficient strategy, "acquaintance immunization," useful when the highest degree nodes are not known [CHb03]; nevertheless, it allows complete immunization of the network with a finite and small threshold. Similar strategies may be applied to the spread of epidemics in human populations and could also be embedded into automated algorithms for network immunization.

3 Observations in real-world networks: the Internet, epidemics, proteins and DNA

During the late 1990s, the advancement and popularity of computers (in particular PCs) changed our understanding of complex networks. The availability and power of computers made it possible to gather huge databases of network structures and to analyze them quickly and efficiently. This allowed, for the first time, the comparison of real network data with the existing models, and, in particular, the ER model. Another significant influence of the computing revolution was the creation of two huge and rapidly developing networks: the Internet and the World Wide Web (WWW).

3.1 Real-world complex networks

Below, we present a few examples of real-world networks, many of which are well approximated by a scale-free degree distribution. Many other examples that exist may be found in the recent literature (see e.g. [AB02, BBV08, BLM$^+$06, DM03, New02b, PV03]).

3.1.1 Computer networks and the Internet

Computer networks are good candidates for studying the role they play in the Internet. The connections in this case are physical – a link exists between two nodes (computers) in the network if they are physically connected by a cable (usually copper or optical fiber, although some other connection types, such as satellites, exist). Various sizes of networks exist, ranging from a simple LAN (local area network) – where usually a small number of computers are all connected together – to wide area networks (WANs), which may consist of tens of thousands of computers.

However, today, computer networks are rarely isolated. Almost all networks, as well as many low-end computers, are connected to the Internet, the network of networks. The Internet consists of many autonomous systems (ASs), i.e., networks that are run by their respective owners. ASs cooperate by exchanging small packets of data among themselves. The packets are transferred between designated computers, called "routers," containing tables that allow them to direct a packet towards its designated target. Most of this work is done using the IP protocol, which is a set of rules for how such packets are treated.

The Internet, as a complex network, is usually studied at two levels: (1) the router level, where each router (or end unit) is a node, and two nodes physically connected are connected by an edge, and (2) the AS level, where each autonomous system is treated as a node, and two ASs that have a physical connection between them are connected by an edge. As of 2009, the AS network consisted of approximately 10^4 nodes; however, several hundreds of thousands of routers and several millions of end units [SS05] exist. Studying the Internet is made somewhat difficult by the fact that much of the Internet backbone (the main "trunk" connections) is commercial, and no complete data on the real structure are publicly available. The existing studies usually use *traceroute* data, i.e., data obtained by tracing the route of packets sent from some source to some destination. However, this may induce some bias in the measurements (see, e.g., [CM05, LBCX03, PR04]). Some new distributed approaches to measuring the Internet attempt to overcome these biases [SS05]. Although these approaches yielded significantly better maps, the scale-free characteristic has not changed.

3.1.2 Technological networks

Other types of technological networks include the electrical power grid, the phone network, and transport networks – roads, airline connections between airports, railroads, and subway networks. Many of these networks tend to be highly correlated with the space in which they reside, i.e., the surface of the earth. Figure 3.1 shows the network of flights of a US airline, where airports are connected if a direct flight exists between them. It can be seen that some airports function as hubs, offering numerous flights to many possible destinations; however, most airports have flights only to one or two nearby destinations [BBPV04].

3.1.3 Virtual technological networks and the WWW

Another type of network is also based on computers and other technologies, but the links, and sometimes also the nodes in these networks, are logical rather than physical. In a sense, they can be seen more as reflecting the underlying social network rather than a physical one. Of these networks, the largest and most widely known is the World Wide Web (WWW), the network of HTML pages that are usually viewed using a browser. Each such page is a node in the network, and if a (hyper) link exists between pages (i.e., pressing the mouse button on some location in a page leads

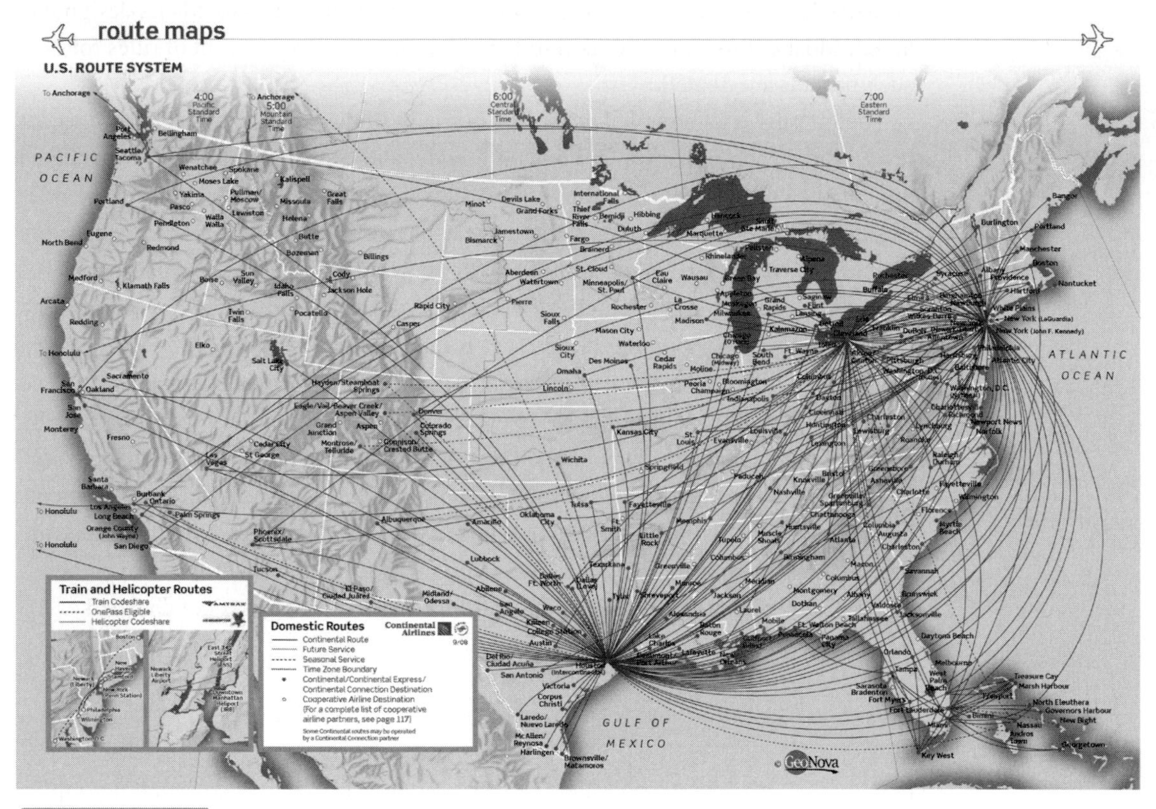

Figure 3.1 Continental Airlines domestic route map.

to the other one), then a directed edge exists between them. This network is huge. In 2009 it was estimated to contain hundreds of billions of pages. The WWW is usually probed and reconstructed using a "robot," or a "crawler," which is a program that automatically scans pages and follows links [AJB99]. Some difficulty exists in defining this network, since many pages today are dynamic, that is, they are created on the fly, based on user data and preferences. Another studied network is the email network, where every individual is a node and is linked to all other individuals in their electronic address book [EMB02].

Another type of logical network is the phone call graph. This is a graph created by phone network operators. Each node represents a phone number, and a directed link exists between nodes if the source of the link initiated a call to the destination within a certain time frame (usually one day) [KOS+07, PBV07, WGHB09]. Yet another type of a typical virtual technological network is the trust network, where a link exists between nodes if one is willing to accept the other's public cryptographic key in order to verify their electronic signature [GGA+02]. Some research was also conducted on the network of function calls in large software projects (where each function is a node and a call from one function to the other defines a directed link) [JKYR06].

3.1.4 Social networks

An important class of networks is the class of networks of social interactions between individuals. These may consist of networks of friendship or acquaintances, working relations, or sexual relations [LEA+01]. Besides their importance to social studies, understanding the structure of these networks is also important to epidemiologists, since these are the networks on which epidemics spread (see Chapter 14 for details).

Some large social networks that have been studied thoroughly are the actor network, where every actor is connected to every other actor with whom he or she has appeared in a movie, and the coauthorship network. In coauthorship and citation networks, a scientist is connected to scientists, with whom she or he has written papers (see Figure 3.2), or to scientists whose work is cited in her/his papers, respectively [dSP65, Red98].

3.1.5 Economic networks

Economic networks can be viewed as a special type of social network. However, the nodes may not represent individuals, but rather companies, countries, or industries. The edges may represent trading relations between companies or countries, companies sharing a director, a stock-holding relationship, correlated fluctuations in stock returns, etc. (see [BCL+04, GAH08, KKK02, OKK04]).

3.1.6 Biological networks

One of the most important and well studied classes of networks are the biological networks. This class contains several different types of network. Biological networks may be logical, representing interactions between proteins, between genes, or between proteins and genes. Interactions between molecules in the cell's metabolic pathways can also be viewed as a network. Although the interactions are physical, links are not physical entities, but rather, the possibility of an interaction between two molecules. Other logical networks are the ecological predator–prey networks (where nodes are species, and a directed link represents predation of one species by the other). Another type of biological network are physical biological networks, such as the nervous

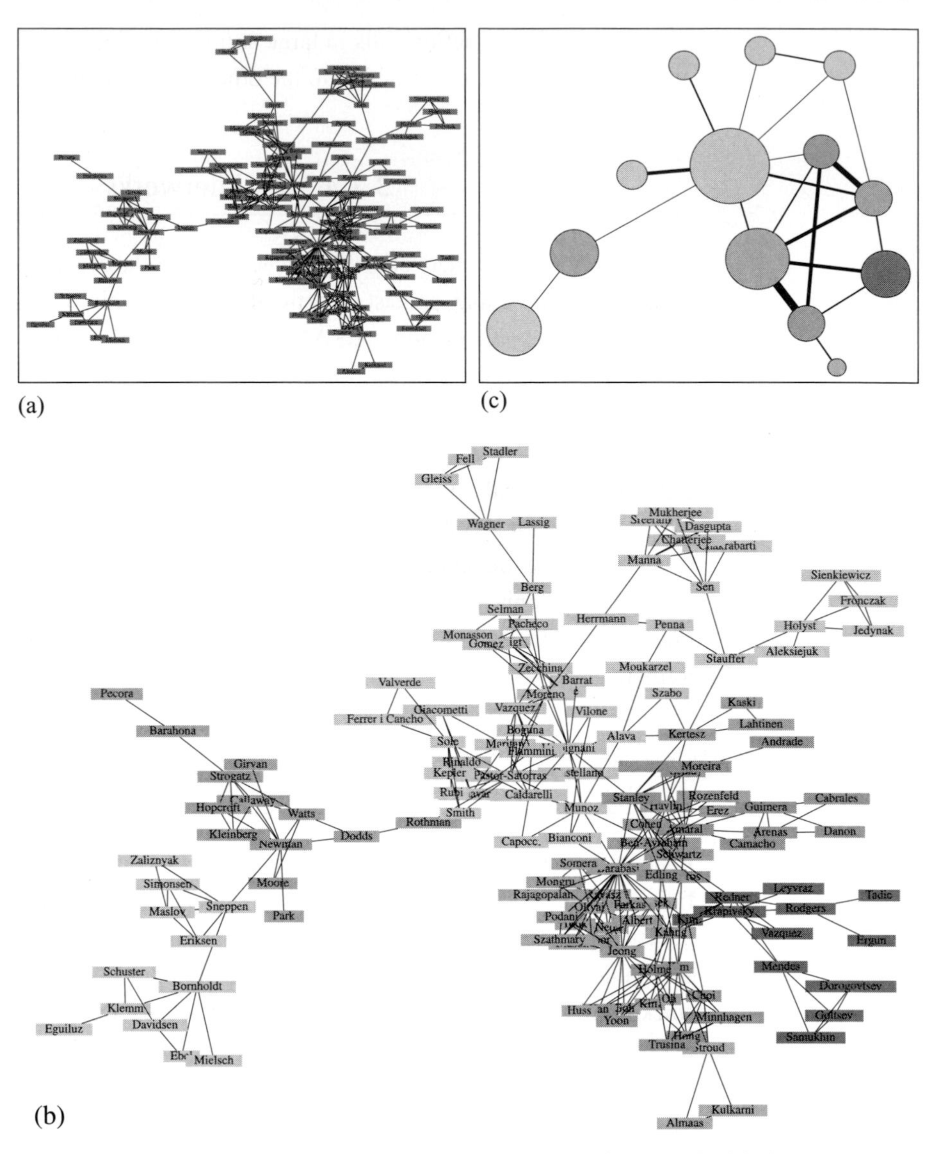

Figure 3.2 The collaboration network of scientists working on network research. (a) The network structure. (b) The network with node names. Each node corresponds to an author, and a link corresponds to coauthorship of at least one paper. Different shades represent different communities identified by a clustering algorithm (see Appendix C for details). The authors of this book can be found in the dark grey group. (c) The different communities identified by the clustering algorithm. After [NG04].

system, the neurons in the brain, and the network of blood vessels in an organism. Several recent studies can be found in [HCE07, MVB$^+$08]. Recently, the structure of neural networks has also been studied and has been shown to be scale free [ECC$^+$05, Kim04]. The network of genes related with known diseases was studied in [GCV$^+$07].

3.1.7 Networks in physics

In the realm of physical phenomena, one may also consider several types of networks, both physical networks, such as solid-state materials, where each atom is linked (interacting) with its nearest neighbor atoms in the lattice, polymer networks [ZKM+03] (where edges connect monomers), and logical networks, such as minima in the energy landscape [Doy02], where each pair of two neighboring local minima are connected, as well as configurations of long chains [SC01], where each node represents a configuration of the chain, and two nodes are connected if one configuration can be reached from the other by a simple maneuver of a monomer. Other physical networks are networks of nano-particles and quantum dots, where links represent hopping between the dots [SHBF05], or optical networks where the nodes are beam splitters and the links are the optical fibers [JKBH08].

3.2 Properties of real-world networks

In Table 3.1 some basic parameters for a number of published networks are presented. The properties measured are as follows: the type of graph, directed or undirected, the total number of vertices n, the total number of edges m, the mean degree z, the mean vertex–vertex distance ℓ, the exponent γ of degree distribution if the distribution follows a power law (or "–" if not; in/out-degree exponents are given for directed graphs), the clustering coefficient $C^{(1)}$ from Eq. (3.2), the clustering coefficient $C^{(2)}$ from Eq. (3.4), and the degree correlation coefficient r (Section 3.2.5). The last column presents the citation(s) for the network in the references. Blank entries indicate unavailable data. In the following subsections, we present detailed definitions of these quantities.

3.2.1 Degree distribution

The degree of a node is the number of links connected to it. In directed networks, one can distinguish between the *in*-degree, *out*-degree, and the total degree (which is the sum of the two). The degree distribution, $P(k)$, is the fraction of sites having degree k. As can be seen above, many real networks do not exhibit a Poisson degree distribution, as predicted in the ER model. In fact, many of them exhibit a distribution with a long, power-law, tail, $P(k) \sim k^{-\gamma}$ with some γ, usually between 2 and 3.

Table 3.1 Basic statistics for a number of published networks

The properties measured are as follows: the type of graph, directed or undirected, the total number of vertices n, the total number of edges m, the mean degree z, the mean vertex–vertex distance l, the exponent γ of degree distribution if the distribution follows a power law (or "–" if not; in/out-degree exponents are given for directed graphs), the clustering coefficient $C^{(1)}$ from Eq. (3.2), the clustering coefficient $C^{(2)}$ from Eq. (3.4), and the degree correlation coefficient r, Section 3.2.5. The last column gives the citation(s) for the network in the references. Blank entries indicate unavailable data. After [New02b].

Network	Type	n	m	z	l	γ	$C^{(1)}$	$C^{(2)}$	r	References
Social										
film actors	undirected	449 913	25 516 482	113.43	3.48	2.3	0.20	0.78	0.208	[ASBS00, WS98]
company directors	undirected	7 673	55 392	14.44	4.60	–	0.59	0.88	0.276	[DYB03, NSW01]
math coauthorship	undirected	253 339	496 489	3.92	7.57	–	0.15	0.34	0.120	[DCG99, GI95]
physics coauthorship	undirected	52 909	245 300	9.27	6.19	–	0.45	0.56	0.363	[New01a, New01b]
biology coauthorship	undirected	1 520 251	11 803 064	15.53	4.92	–	0.088	0.60	0.127	[New01a, New01b]
telephone call graph	undirected	47 000 000	80 000 000	3.16		2.1				[ACL00, ACL01]
email messages	directed	59 912	86 300	1.44	4.95	1.5/2.0		0.16		[EMB02]
email address books	directed	16 881	57 029	3.38	5.22	–	0.17	0.13	0.092	[NFB02]
student relationships	undirected	573	477	1.66	16.01	–	0.005	0.001	−0.029	[BMS04]
sexual contacts	undirected	2 810				3.2				[LEA+01, LEA03]
Information										
WWW nd.edu	directed	269 504	1 497 135	5.55	11.27	2.1/2.4	0.11	0.29	−0.067	[AJB99, BAJ00]
WWW Altavista	directed	203 549 046	2 130 000 000	10.46	16.18	2.1/2.7				[BKM+00]
citation network	directed	783 339	6 716 198	8.57		3.0/–				[Red98]
Roget's Thesaurus	directed	1 022	5 103	4.99	4.87	–	0.13	0.15	0.157	[Knu93]
word co-occurrence	undirected	460 902	17 000 000	70.13		2.7		0.44		[CS01, DM01b]

Technological										
Internet	undirected	10 697	31 992	5.98	3.31	2.5	0.035	0.39	−0.189	[CCGJ02, FFF99]
power grid	undirected	4 941	6 594	2.67	18.99	–	0.10	0.080	−0.003	[WS98]
train routes	undirected	587	19 603	66.79	2.16	–		0.69	−0.033	[SDC+03]
software packages	directed	1 439	1 723	1.20	2.42	1.6/1.4	0.070	0.082	−0.016	[New03b]
software classes	directed	1 377	2 213	1.61	1.51	–	0.033	0.012	−0.119	[VCS02]
electronic circuits	undirected	24 097	53 248	4.34	11.05	3.0	0.010	0.030	−0.154	[FiCJS01]
peer-to-peer network	undirected	880	1 296	1.47	4.28	2.1	0.012	0.011	−0.366	[ALPH01, RFI02]
Biological										
metabolic network	undirected	765	3 686	9.64	2.56	2.2	0.090	0.67	−0.240	[JTA+00]
protein interactions	undirected	2 115	2 240	2.12	6.80	2.4	0.072	0.071	−0.156	[JMBO01]
marine food web	directed	135	598	4.43	2.05	–	0.16	0.23	−0.263	[HBR96]
freshwater food web	directed	92	997	10.84	1.90	–	0.20	0.087	−0.326	[Mar91]
neural network	directed	307	2 359	7.68	3.97	–	0.18	0.28	−0.226	[WS98, WSTB86]

3.2.2 Distances and optimal paths

Since many networks are not embedded in real space, the geometrical distance between nodes is meaningless. The most important distance measure in such networks is the minimal number of hops (or chemical distance). That is, the distance between two nodes in the network is defined as the number of edges in the shortest path between them. If the edges are assumed to be weighted, the lowest total weight path, called the *optimal path*, may also be used. The usual mathematical definition of the diameter of the network is the length of the path between the farthest nodes in the network.

3.2.3 Edge lengths and geography

Measuring the Internet reveals that the length of the physical cables connecting nodes, r, is distributed according to [YJB02]:

$$P(r) \sim r^{-1}, \quad r \leq r_{\max} \,. \tag{3.1}$$

This is in contrast to the number of nodes at a distance r from a given node that is approximately proportional to r, since they are embedded in a real two-dimensional surface (in fact, there are some indications that the distribution of population density is fractal, but it still has a dimension between 1 and 2). Thus, if the nodes were connected randomly, with no geography-related bias, we would expect to see the number of links increase with the length. This indicates that distant nodes have a much lower probability of being connected to each other than nearby nodes. This property cannot be captured by the ER model, which does not have any geographical distances. A discussion of several possible geographical models that account for real distances will be presented in Sections 4.2 and 4.6. For a recent study taking into account geographical constraints see [KHB08].

3.2.4 Clustering

The clustering coefficient is usually related to a community represented by local structures. The usual definition of clustering (sometimes also referred to as transitivity) is related to the number of triangles in the network. The clustering is high if two nodes sharing a neighbor have a high probability of being connected to each other.

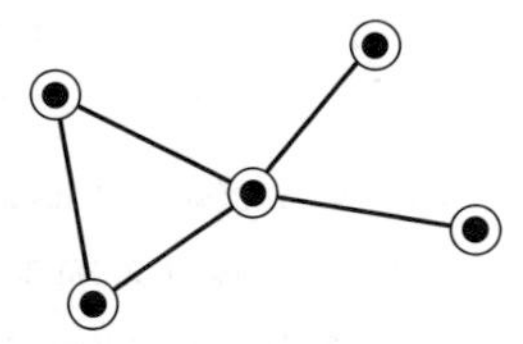

Figure 3.3 Illustration of the definition of the clustering coefficient C, Eqs. (3.2) and (3.4). This network has one triangle and eight connected triples, and therefore has a clustering coefficient of $3 \times 1/8 = 3/8$ according to Eq. (3.2). The individual vertices have local clustering coefficients, Eq. (3.3), of 1, 1, $\frac{1}{6}$, 0 and 0, for a mean value, Eq. (3.4), of $C = 13/30$. After [New02b].

There are two common definitions of clustering. The first is global,

$$C = \frac{3 \times \text{ the number of triangles in the network}}{\text{the number of connected triples of vertices}}, \tag{3.2}$$

where a "connected triple" means a single vertex with edges running to an unordered pair of other vertices (see Figure 3.3).

A second definition of clustering is based on the average of the clustering for single nodes. The clustering for a single node is the fraction of pairs of its linked neighbors out of the total number of pairs of its neighbors:

$$C_i = \frac{\text{the number of triangles connected to vertex } i}{\text{the number of triples centered on vertex } i}. \tag{3.3}$$

For vertices with degree 0 or 1, for which both numerator and denominator are zero, we use $C_i = 0$. Then the clustering coefficient for the whole network is the average

$$C = \frac{1}{n} \sum_i C_i. \tag{3.4}$$

In both cases, the clustering is in the range $0 \le C \le 1$.

In random graph models such as the ER model and the configuration model (presented in Section 4.3), the clustering coefficient is low and decreases to 0 as the system size increases. This is also the situation in many growing network models. However, in many real-world networks the clustering coefficient is rather high and remains constant for large network sizes (see Table 3.1). This observation led to the introduction of the small-world model, presented in Section 4.2, which offers a combination of a regular lattice with high clustering and a random graph. Another approach, based on the community structure, is presented in [GN02].

3.2.5 Correlations

In random graph models, it is usually assumed that there are no correlations between the degrees of neighboring nodes. That is, the probability of reaching a node by following a link is independent of the node from which the link emanated. In many real-world networks, however, this is not the case. Several types of correlations exist, depending on the internal properties of the nodes. However, when considering only the network topology, the main types of correlations that have been studied are the degree-degree correlations.

Degree-degree correlations are represented by $P(k, k')$, the probability that a node of degree k is connected to a node of degree k'. If no correlation exists, given an edge, then the probability that it leads to a node of degree k is $kP(k)/\langle k\rangle$. Thus, the probability that an edge leads from a node of degree k to a node of degree k' is $P(k, k') = kk'P(k)P(k')/\langle k\rangle^2$ (where each direction of the edge is counted separately). If correlations exist, the joint probability will deviate from this form [KR01]. An alternative approach for studying correlations is analyzing the average degree of neighboring nodes as a function of the degree, i.e., $\langle k\rangle_{nn}(k)$. This yields a one-parameter curve that can be easily studied. One can also calculate the correlation coefficient, r, between the degrees of neighboring sites [New02a]:

$$r = \frac{\langle k_i k_j\rangle - \langle k\rangle^2}{\langle k^2\rangle - \langle k\rangle^2}, \tag{3.5}$$

where averages are taken over all pairs of neighbors, i and j. Another method of dealing with correlations is presented in [GSM08], where the joint probability of a link between two nodes having degrees k_1 and $k_2 < k_1$ is approximated by $P(k_1, k_2) \sim k_1^{-(\gamma-1)} k_2^{-\epsilon}$ and the exponent ϵ is used to signify the correlations.

3.2.6 Modular and hierarchical networks

Some networks are composed of several different modules usually associated with different groups of nodes or different functions of the modules in the overall functioning of the network. Many biological networks, such as genetic and metabolic networks, have a modular structure [RSM$^+$02]. This is attributed to the different groups of genes or metabolites that are responsible for different functions of the cell. These modules can usually be distinguished by the fact that nodes in the module are strongly connected – modules tend to be densely connected internally. However, modules are relatively weakly interconnected. Appendix C presents a discussion of different methods for studying the modular structure of a network.

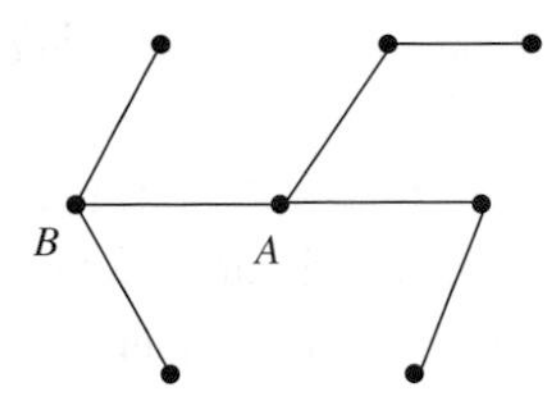

Figure 3.4 The betweenness centrality of node A is higher than that of node B, although both have the same degree.

A more implicit form of modularity exists in networks with hierarchical structure. These networks are called hierarchical networks [DGM02a, RSM+02], since they are usually constructed from a hierarchy of modules. Modules at the high levels, which are usually related to some high level function, are composed of smaller modules related to different subfunctions of this high level function. Models of such networks are usually constructed by a recursive process. A further generalization of this idea is the notion of fractal, self-similar, networks [SHM05], where the structure of the network becomes fractal-like. For a further discussion, see Chapter 7.

3.3 Betweenness centrality: what is your importance in the network?

The importance of a node in a network depends on many factors. A website may be important due to its content, a router due to its capacity, a metabolite due to its biochemical function, etc. Of course, all of these properties depend on the nature of the studied network, and may have very little to do with the graph structure of the network. Since the subject of this book is networks, we are particularly interested in the importance of a node (or a link) due to its topological function in the network.

It is reasonable to assume that the topology of a network may dictate some intrinsic importance for different nodes. One measure of centrality can be the degree of a node. The higher the degree, the more the node is connected, and therefore, the higher is its centrality in the network. However, the degree is not the only factor determining a node's importance. In examining Figure 3.4 for instance, it seems obvious that node A is more central than node B, although both have the same degree. Possible alternative definitions may be how removing the node influences the number of nodes that are still connected in the network, or how this influences the average distance from a node to all other nodes in the network. Both these definitions have distinct disadvantages. Removing a central node may have little influence on the network, if the entire network is well connected, whereas removing a peripheral node may disconnect some of its neighbors, which may themselves be unimportant and peripheral. A "parasitic" node may have a small average distance, compared with the rest of the network, merely by being connected to a very central node, whereas the "parasite" itself has no influence on the function of the network.

One of the most accepted definitions of centrality is based on counting paths going through a node. For each node, i, in the network, the number of "routing" paths to all other nodes (i.e., paths through which data flow) going through i is counted, and this number determines the centrality i. The most common selection is taking only the shortest paths as the routing paths. This leads to the following definition: the *betweenness centrality* of a node, i, equals the number of shortest paths between all pairs of nodes in the network going through it, i.e.,

$$g(i) = \sum_{\{j,k\}} g_i(j,k) , \tag{3.6}$$

where the notation $\{j, k\}$ stands for summing each pair once, ignoring the order, and $g_i(j, k)$ equals 1 if the shortest path between nodes j and k passes through node i and 0 otherwise. In fact, in networks with no weight (i.e., where all edges have the same length), there might be more than one shortest path. In that case, it is common to take $g_i(j, k) = C_i(j, k)/C(j, k)$, where $C(j, k)$ is the number of shortest paths between j and k, and $C_i(j, k)$ is the number of those going through i. Several variations of this scheme exist, focusing, in particular, on how to count distinct shortest paths (if several shortest paths share some edges). These differences tend to have a very small statistical influence in random complex networks, where the number of short loops is small. Therefore, we will concentrate on the above definition (Eq. (3.6)) here. Another nuance is whether the source and destination are considered part of the shortest path. This is also irrelevant for very high degree nodes, on which we will mainly focus.

Figure 3.4 shows that 12 shortest paths pass through node A (not counting paths originating or ending at A), whereas 10 pass through B. This indicates that A is more central in the network, as our intuition indicated. Since the network in the figure is a tree, there is only one shortest path between any pair of nodes. Moreover, another indication of the centrality of A is that removing A will break the network into three small networks, whereas removing B only disconnects two other nodes from the network. The usefulness of the betweenness centrality in identifying bottlenecks and important nodes in the network has led to applications in identifying communities in biological and social networks [GN02].

In [GKK01b, GOJ$^+$02] the behavior of the distribution of betweenness, $P_B(g)$, in real and model networks was studied. It was found that for these networks the distribution follows a power law, $P_B(g) \sim g^{-\eta}$. This is in contrast to ER networks, where the betweenness is very homogeneous (as expected by the homogeneous

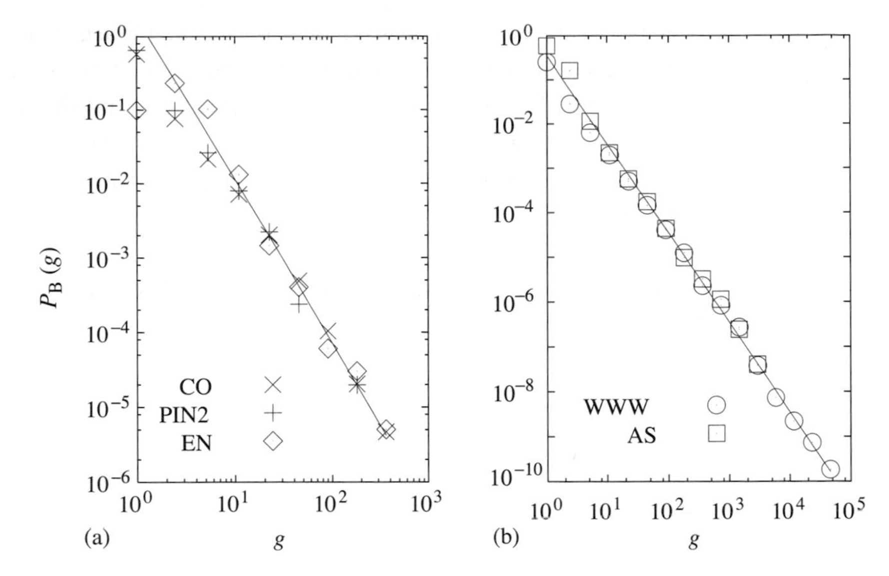

Figure 3.5 Betweenness centrality (BC) distribution for several real networks. (a) The distribution of BC is presented for the protein interaction network of the worm *S. cerevisiae* (PIN2), the coauthorship network in the field of neuroscience (CO), and the metabolic network of a eukaryotic organism, *Emericella nidulans* (EN). (b) The distribution of BC is presented for the WWW and the Internet at the AS level. After [GOJ+02].

degree distribution) and $P_B(g)$ decays exponentially. In fact, for most real-world networks, it was found that $\eta \approx 2$, whereas for artificial networks and models having a scale-free degree distribution, $\eta \approx 2.2$. In [Bar04] this universality is disputed. However, it is agreed that the latter exhibits a power-law behavior with η a little higher than 2. Figure 3.5 presents this distribution for several real-world networks. For further results, see [AB02].

There are other approaches to the importance of nodes. A well-known example is the Page Rank algorithm [PBMW99] used to determine the importance of WWW pages based on the links pointing to them. This algorithm initiates a random walk at a random node, following a random link at each node, with some small probability, at every step, of jumping to a randomly chosen node without following a link. This algorithm gives high importance (high probability of hitting) to nodes with a high number of links pointing to them, and also to nodes pointed to by these nodes.

3.4 Conclusions

As seen from the measurements presented above, for almost all studied networks, many properties cannot be explained by the ER model. However, this only became apparent in recent years when methods for studying real networks were developed. The failure of the simple model to explain the behavior of real networks led to the development of new models that account for such properties. The following chapters will cover several models for random networks in attempting to reproduce at least some of the main properties of real-world networks, in an effort to understand them better.

4 Models for complex networks

4.1 Introduction

In this chapter and the following one, we will discuss models for generating complex networks. These models describe the process of creating complex networks and thus also define probability spaces of random graphs. As expected, different methods of creation can lead to different classes of random graphs having completely different properties. We will discuss the properties of networks formed by some of these methods in the second part of this book. This chapter will mainly focus on static models, that is, models for building a complex network with a given set of nodes, and some wiring process (for connecting nodes by links), depending on the desired properties of the network.

It should be clarified that real networks are not random. Their formation and development are dictated by a combination of many different processes and influences. These influencing conditions include natural limitations and processes, human considerations such as optimal performance and robustness, economic considerations, natural selection and many others. Controversies still exist regarding the measure to which random models represent real-world networks. However, in this book we will focus on random network models and attempt to show how their properties may still be used to study properties of real-world networks.

In the next chapter, we will discuss dynamical models of growing networks, where the network starts from a small seed and grows with time by adding new nodes and links. This class of models attempts not only to construct networks with some desired property, but also to explain the manner by which networks are formed in the real world, and to predict expected properties of real-world networks.

4.2 Introducing shortcuts: small-world networks

4.2.1 Introduction

As we have seen in Chapter 3, many real-world networks have many properties that cannot be explained by the ER model. One such property is the high clustering

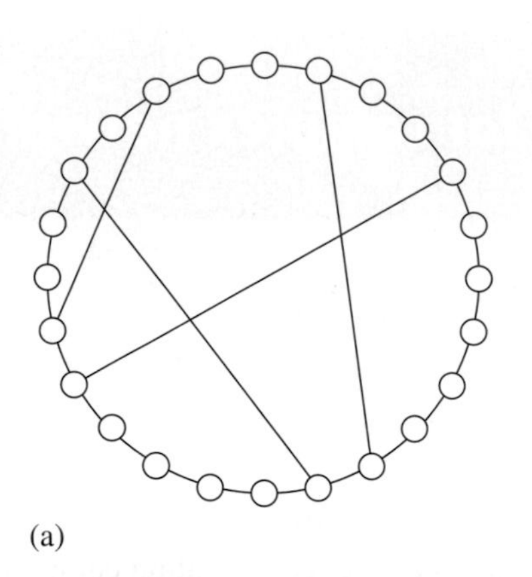

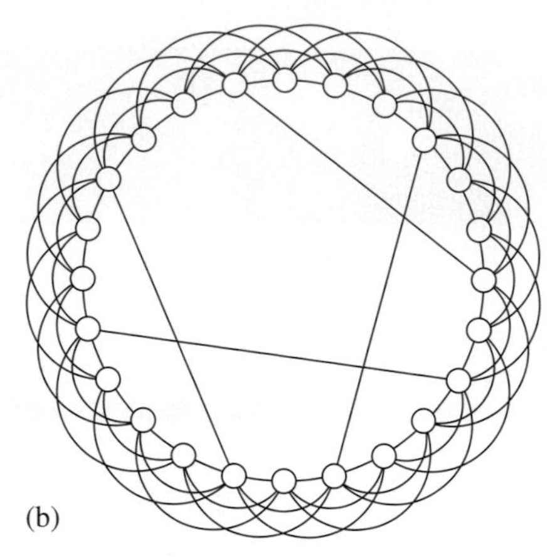

Figure 4.1 Small-world networks with an underlying ring structure. (a) A small-world network formed by adding shortcuts to a simple ring (with $k = 1$). (b) A small-world network formed by adding shortcuts to a ring with $k = 3$. After [NW99].

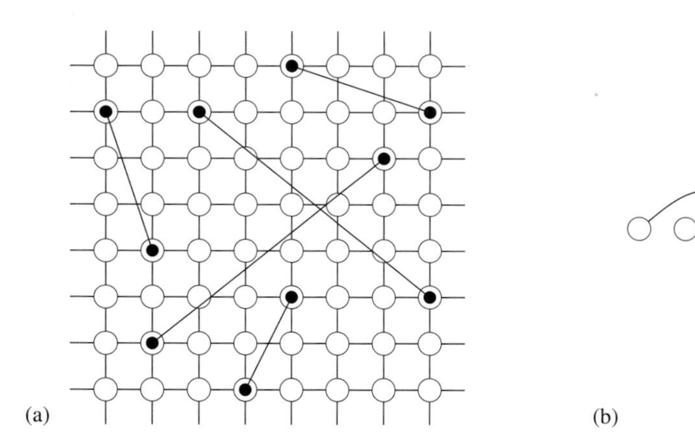

Figure 4.2 An example of a small-world network with an underlying structure of a two-dimensional lattice. (a) The network obtained from adding shortcuts to a two-dimensional grid. (b) The nearest neighbor structure of a node in a two-dimensional lattice with $k = 3$ nearest neighbors on each side (in this example, it was assumed that diagonal neighbors will not be connected). After [NW99].

observed in many real-world networks. This led Watts and Strogatz to develop an alternative model, called the "small-world" model [WS98]. Their idea was to begin with an ordered lattice, such as the k-ring (a ring where each site is connected to its $2k$ nearest neighbors – k from each side) or the two-dimensional lattice (see Figures 4.1 and 4.2). For each site, each of the links emanating from it is removed with probability φ and is rewired to a randomly selected site in the network. A variant of this process is to add links rather than rewire, which simplifies the analysis without considerably affecting the results. The obtained network has the desirable properties of both an ordered lattice (large clustering) and a random network (small world), as we will discuss below.

4.2.2 Clustering in a small-world network

The simplest way to treat clustering analytically in a small-world network is to use the link addition, rather than the rewiring model. In the limit of large network size, $N \to \infty$, and for a fixed fraction of shortcuts, φ, it is clear that the probability of forming a triangle vanishes as we approach $1/N$, so the contribution of the shortcuts to the clustering is negligible. Therefore, the clustering of a small-world network is determined by its underlying ordered lattice. For example, consider a ring where each node is connected to its k closest neighbors from each side. A node's number of neighbors is therefore $2k$, and thus it has $2k(2k-1)/2 = k(2k-1)$ pairs of neighbors. Consider a node, i. All of the k nearest nodes on i's left are connected to each other, and the same is true for the nodes on i's right. This amounts to $2k(k-1)/2 = k(k-1)$ pairs. Now consider a node located d places to the left of k. It is also connected to its k nearest neighbors from each side. Therefore, it will be connected to $k-d$ neighbors on i's right side. The total number of connected neighbor pairs is

$$k(k-1) + \sum_{d=1}^{k}(k-d) = k(k-1) + \frac{k(k-1)}{2} = \frac{3}{2}k(k-1)\,, \tag{4.1}$$

and the clustering coefficient is:

$$C = \frac{\frac{3}{2}k(k-1)}{k(2k-1)} = \frac{3(k-1)}{2(2k-1)}\,. \tag{4.2}$$

For every $k > 1$, this results in a constant larger than 0, indicating that the clustering of a small-world network does not vanish for large networks. For large values of k, the clustering coefficient approaches $3/4$, that is, the clustering is very high.

Note that for a regular two-dimensional grid, the clustering by definition is zero, since no triangles exist. However, it is clear that the grid has a neighborhood structure. This led to several definitions in which it was attempted to replace the definition of clustering as the fraction of triangles by alternative definitions [SKO+07]. However, the standard definition has remained the most commonly used one.

4.2.3 Distances in a small-world network

The second important property of small-world networks is their small diameter, i.e., the small distance between nodes in the network. The distance in the underlying lattice behaves as the linear length of the lattice, L. Since $N \sim L^d$, where d is the lattice dimension, it follows that the distance between nodes behaves as:

$$l \sim L \sim N^{1/d} . \quad (4.3)$$

Therefore, the underlying lattice has a finite dimension, and the distances on it behave as a power law of the number of nodes, i.e., the distance between nodes is large. However, when adding even a small fraction of shortcuts to the network, this behavior changes dramatically.

To deduce the behavior of the average distance between nodes, we follow [NW99]. Consider a small-world network, with dimension d and connecting distance k (i.e., every node is connected to any other node whose distance from it in every linear dimension is at most k[1]). Now, consider the nodes reachable from a source node with at most r steps. When r is small, these are just the rth nearest neighbors of the source in the underlying lattice. We term the set of these neighbors a "patch," the radius of which is kr, and the number of nodes it contains is approximately $n(r) = (2kr)^d$.

We now want to find the distance r for which such a patch will contain about one shortcut. This will allow us to consider this patch as if it was a single node in a randomly connected network. Assume that the probability for a single node to have a shortcut is ϕ.[2] To find the length for which approximately one shortcut is encountered, we need to solve for r the following equation: $(2kr)^d \phi = 1$. The correlation length, ξ, defined as the distance (or linear size of a patch) for which a shortcut will be encountered with high probability is therefore,

$$\xi = \frac{1}{k\phi^{1/d}} . \quad (4.4)$$

Note that we have omitted the factor 2, since we are interested in the order of magnitude.

Let us denote by $V(r)$ the total number of nodes reachable from a node by at most r steps, and by $a(r)$, the number of nodes added to a patch in the rth step. That is, $a(r) = n(r) - n(r-1)$. Thus,

$$a(r) \approx \frac{dn(r)}{dr} = 2kd(2kr)^{d-1} . \quad (4.5)$$

When a shortcut is encountered at the r' step from a node, it leads to a new patch.[3] This new patch occurs after r' steps, and therefore the number of nodes reachable

[1] Note that in one dimension this is the same as the definition of k above.
[2] Note that ϕ is not exactly the same as φ defined above, since φ is defined as the probability to rewire (or add) every link as a shortcut, whereas ϕ is the probability per node.
[3] It may actually lead to an already encountered patch, and two patches may also merge after some steps, but this occurs with negligible probability when $N \to \infty$ until most of the network is reachable.

from its origin is $V(r - r')$. Thus, we obtain the recursive relation,

$$V(r) = \sum_{r'=0}^{r} a(r') \left[1 + \xi^{-d} V(r - r')\right] , \tag{4.6}$$

where the first term stands for the size of the original patch, and the second term is derived from the probability of hitting a shortcut, which is approximately ξ^{-d} for every new node encountered.

To simplify the solution of Eq. (4.6), it can be approximated by a differential equation. The sum can be approximated by an integral, and then the equation can be differentiated with respect to r. For simplicity, we will concentrate here on the solution for the one-dimensional case, with $k = 1$, where $a(r) = 2$. Thus, one obtains

$$\frac{\mathrm{d}V(r)}{\mathrm{d}r} = 2\left[1 + V(r)/\xi\right] , \tag{4.7}$$

the solution of which is:

$$V(r) = \xi(\mathrm{e}^{2r/\xi} - 1) . \tag{4.8}$$

For $r \ll \xi$, the exponent can be expanded in a power series, and one obtains $V(r) \approx 2r = n(r)$, as expected, since usually no shortcut is encountered. For $r \gg \xi$, $V(r)$ grows exponentially as in a random graph. An approximation for the average distance between nodes can be obtained by equating $V(r)$ from Eq. (4.8) to the total number of nodes, $V(r) = N$. This results in

$$r \approx \frac{\xi}{2} \ln \frac{N}{\xi} . \tag{4.9}$$

As apparent from Eq. (4.9), the average distance in a small-world network behaves as the distance in a random graph with patches of size ξ behaving as the nodes of the random graph.

4.3 Random graphs with a given degree distribution

4.3.1 Introduction

In Chapter 2, we discussed the Erdős–Rényi models for random networks. In these models the edges are randomly distributed between nodes. As mentioned there, this leads to a Poisson distribution of the degrees and to the small-world property in which the distance is proportional to $\log N$ (where N is the number of nodes). However, as shown in Chapter 3, the degree distribution in real networks is often very broad and far from Poissonian. In this chapter we discuss the "equilibrium" ensemble of networks with a prescribed degree distribution, i.e., the ensemble gives equal representation to all networks having a given degree distribution. In Chapter 5 we will discuss growing models, the aim of which is to simulate the growth and development of real networks. We will see that these models generate ensembles that are not necessarily identical to the equilibrium ensemble. However, since many real-world networks seem quite random, and in many cases there is no knowledge of the real processes driving them, it seems natural, as a first step, to study the equilibrium ensemble and its properties. This will be our main focus in this book.

4.3.2 Random regular graphs

The first approach that proposed to generate random networks with a given degree distribution was the model of random regular networks. The ensemble includes all networks having N nodes and m links per node (of course Nm must be even). Bollobás [Bol80] suggested a method for producing all such graphs with the same probability. This method is based on putting Nm balls in an urn, m copies for each node i, and with randomly matching pairs connecting the respective nodes. More formally:

(i) produce m copies of each node (representing "open" links);
(ii) randomly select pairs of open links and connect the corresponding nodes.

This method may lead, of course, to a multigraph, either by connecting a node to itself or by connecting two nodes together more than once. However, as N increases, these events become rare and their number becomes statistically insignificant.

The ensemble of random m-regular graphs for $m \geq 3$ has some nice features[4] that make it interesting for study. The most important one is that it is fully connected, i.e., since $N \to \infty$ a path exists between every two nodes (or, to state it differently, all nodes belong to the same component). This also makes it easier to calculate average properties over nodes since we do not need to base the averaging on the fact that a path exists between them.

To understand the structure of such networks, note that the number of neighbors of every node is m, the number of second neighbors is $m(m-1)$, and in general, the number of lth neighbors is:

$$N_l = m(m-1)^{l-1} . \tag{4.10}$$

This is true as long as $(m-1)^l \ll N$, since in this regime the number of nodes reached is much smaller than N, and therefore the probability of reaching the same node twice

[4] For $m = 1$ the model only yields a random pairing of the nodes, and for $m = 2$, it consists of several disconnected cycles.

is statistically insignificant. This implies that the distance between nodes in random regular networks, D, can be obtained approximately by requiring $m(m-1)^{D-1} \approx N$ (since most nodes are reached, and every node has at least a high probability of connecting to a reached node), resulting in

$$D \approx \frac{\log N}{\log(m-1)} . \tag{4.11}$$

Other consequences of these considerations are the connectedness property mentioned above, and the expansion of the graph. The expansion of a graph has several definitions. We will discuss vertex expansion, which is the ratio of the number of neighbors of a group of nodes to the number of nodes in the group. In random m-regular graphs every set of $n \ll N$ nodes has at least $(m-1)n$ neighbors. This means intuitively that there are no "bottlenecks" in such networks, and therefore the network has desirable properties in terms of transmitting data efficiently, mixing of random walks, and is also suitable for other applications, such as designing error correcting codes [SS96]. A more detailed discussion can be found in Chapter 17.

4.3.3 Generalized random graphs

The model for network generation presented above has also been extended in [Bol80] to other degree distributions. One can either fix the degree of each node, i, to be k_i, or, alternatively, choose the degree of each node from some distribution, $P(k)$. The regular graph model can be viewed as a special case of this model, with $k_i = m$ for every node, or with $P(k) = \delta_{k,m}$. The ER model is also very similar to the model with the Poisson degree distribution, $P(k) = \mathrm{e}^{-c}c^k/k!$ (Eq. (2.1)), where $c = \langle k \rangle$ is the average degree. A more formal description of this model is:

(i) for each i, randomly choose k_i from the distribution $P(k)$;
(ii) produce k_i copies of each node, i (representing "open" links);
(iii) randomly select pairs of open links and connect the corresponding nodes (while removing the copies from the open links list).

Figure 4.3 illustrates the different stages of this algorithm. This model has been shown to produce every graph within the given degree sequence with equal probability[5] [Bol80]. In the rest of this book, we will mostly discuss the properties of graphs generated by this approach.

[5] There is actually a requirement that the degrees should not be too large (at most $N^{1/4}$). However, when treating the configuration model as the *definition* of the ensemble, larger degrees can also be considered up to $N^{1/2}$ (see also [PSS05]).

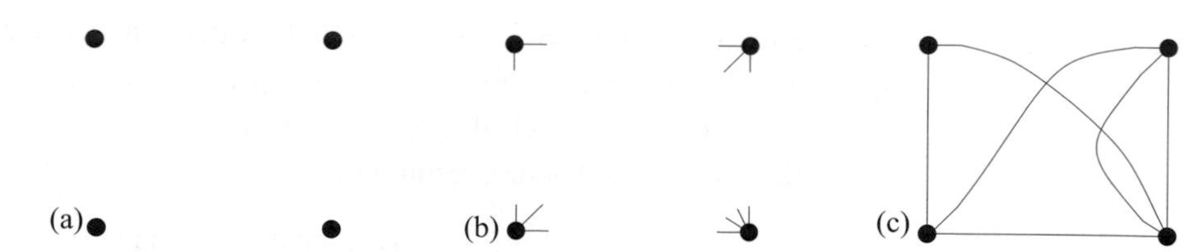

Figure 4.3 Illustration of the stages of creating the generalized random graph. (a) The N ($N = 4$) nodes are created. (b) The degree of each of the nodes is selected and k_i "stubs" ("open links") are produced for each node, i. (c) The stubs are randomly matched. Note that the graph produced here is a multigraph: the two rightmost nodes are doubly connected.

4.3.4 A related model

In physical, thermodynamic, and transport systems, it is usually easier to study the properties of the **canonical ensemble** rather than the **microcanonical ensemble**. This stems from the fact that the second one requires averaging only over microscopic states with the given energy, whereas the first one has a constant temperature, T, rather than energy, E. This allows averaging over all microscopic states with a weight function $e^{-E/k_B T}$ known as the **Boltzmann factor**. This summation (or integration) is usually easier since it does not induce "hard" constraints on the variables. Similarly, in ER networks, calculating in the model $G_{N,p}$ (see Chapter 2 for definitions) is usually easier than calculating in the model $G_{N,M}$, since in the latter, there are non-trivial correlations between different links. Having a link at a certain location has a small but existent effect on the probability of having a link somewhere else, owing to the conservation of the total number of links. Since the effect is small for almost every p except for the extreme cases of almost empty or almost complete graphs,[6] usually problems are solved in the context of the $G(N, p)$ model and the results are also valid for the appropriate $G(N, M)$ model, having $M = Np$.

Such a model also exists in the context of scale-free graphs or generalized random graphs in general. The model in [ACL00] consists of the following construction:

- for each node, i, select k_i from the distribution $P(k)$;
- for each pair of nodes, i and j, connect i and j with probability $P(i \leftrightarrow j) = k_i k_j / N$.

Note that this method only applies if the upper cutoff of $P(k)$ is at most $\sqrt{N}$, otherwise $P(i \leftrightarrow j)$ may exceed one. Also, note that here k_i plays the role of the *average* degree of a node, rather than the actual degree, since a node may end up having a different number of neighbors than its assigned k_i. The advantage of this model is in avoiding the dependence between links of a node, since $P(i \leftrightarrow j)$ is independent of $P(i \leftrightarrow l)$ for every $l \neq j$, when their k are known. This is mainly helpful when proving rigorous

[6] For proof and restrictions, see [Bol85].

results regarding this class, since in calculations we tend to ignore this dependence because it is weak and has a negligible influence on the results.

4.4 Introducing correlations

A final static model discussed here is networks with a given degree sequence and correlations. Correlations in real networks were discussed in Section 3.2.5. The most general form of degree-degree correlations is obtained by having a probability density function $P(k_1, k_2)$, which is the probability of having a link between a node of degree k_1 and a node of degree k_2. In order to be consistent, $P(k_1, k_2)$ should be symmetric and k_1 and k_2 (for undirected graphs) should satisfy the equality $\sum_{k_2} P(k_1, k_2) = P(k_1)k_1/N$. Equivalently, one may use the conditional probability density $P(k_2|k_1)$, which is the probability of reaching a node having degree k_2 by a link emanating from a node of a given degree k_1.

Producing correlations using a completely static model similar to the one presented in Section 4.3 is difficult since not only the degree of a node, but also its probability of connecting to each neighbor should be taken into account. The method usually used to generate such networks is by shuffling links using some sort of a Metropolis-type algorithm [MRR$^+$53]. More details are given in Appendix C.

4.5 Randomly directed networks: modeling the WWW

4.5.1 Introduction

The World Wide Web (WWW) is a collection of webpages stored on many servers around the world. These pages are accessed using a communication protocol known as HTTP (Hyper Text Transfer Protocol). This protocol dictates the details of the request sent by the client (a user's computer) for a certain page or resource, and the form of the reply by the web server. The pages themselves are written using a standard language called HTML (Hyper Text Markup Language). This language allows the introduction of links to other pages in the same site or in remote sites, which are traversed by clicking the mouse on them. Each linked page is identified by its URL (Unified Resource Locator), which is a text string containing the site's address and the path to the page on the site. Of course, the WWW is much more complicated than is presented here. Many webpages do not exist anywhere, but are created on the fly upon user requests by database queries and other methods. However, as a first approximation, this simple model can explain the structure of the WWW.

To model this structure, we can view the webpages as nodes in a directed graph, and every link as a directed edge in the network, pointing from the referring page to the linked page. Other examples of directed networks are biochemical networks, where nodes are the different materials (e.g. proteins) and directed edges indicate that one material catalyzes the generation of another. In addition, in many non-directed networks, such as the Internet, road and airline networks as well as others, there are unidirectional links that cannot be traversed in the opposite direction. Additional examples are satellite links in communication networks, which are usually directed, and unidirectional vehicular roads.

4.5.2 Generalized randomly directed graphs

The model for generalized randomly directed networks is very similar to that of the non-directed networks presented in Section 4.3. The difference is that here we have a joint degree distribution $P(j,k)$, which is the probability of a node to have j links as in-degree (i.e., the number of incoming links) and k links as out-degree. Alternatively, when there are no correlations between the in- and out-degrees, two degree distributions can be given, $P_{\rm in}(k)$ for the in-degree and $P_{\rm out}(k)$ for the out-degree. To allow the actual construction of such networks, these distributions must obey the relation:

$$\sum_{j,k} jP(j,k) = \sum_{j,k} kP(j,k)\,, \tag{4.12}$$

or

$$\sum_{k} kP_{\rm in}(k) = \sum_{k} kP_{\rm out}(k)\,. \tag{4.13}$$

This guarantees that the total number of incoming links is identical (at least statistically) to the number of outgoing links, which is necessary because every link must leave one node and enter the other. The full generation procedure can be stated as follows.

(i) For each node, i, randomly choose j_i and k_i from the distribution $P(j,k)$.
(ii) Produce j_i copies of each node, i, in the "in" list and k_i copies of each node, i, in the "out" list.
(iii) Randomly select an open link from the "in" list and one from the "out" list and connect the corresponding sites by a directed link (going from the in to the out node).

For the structural properties of such directed graphs, see Chapter 11.

4.6 Introducing geography: embedded scale-free lattices

In this section we describe methods for embedding scale-free networks, with degree distribution $P(k) \sim k^{-\gamma}$, in regular Euclidean lattices. The embedding is driven by a natural constraint of minimization of the total (Euclidean) length of the links in the system. We show that all networks with $\gamma > 2$ can be successfully embedded up to a (Euclidean) distance ξ, which can be made as large as desired upon changing an external parameter. Clusters of successive chemical shells (i.e., layers of nodes at the same hop distance from a given node) are found to be compact (the fractal dimension is $d_f = d$), whereas the dimension of the shortest path between any two nodes is smaller than one: $d_{\min} = \frac{\gamma-2}{\gamma-1-1/d}$, contrary to all other known examples of fractals and disordered lattices, where $d_{\min} \geq 1$ [BH94, Hb87].

4.6.1 Introduction

As shown in Chapter 3, many social, biological, and communication systems can be properly described by complex networks whose nodes represent individuals or organizations and whose links mimic the interactions among them [AB02, BBV08, BS02, DM02, DM03, PV03]. An important class of complex networks are the *scale-free* networks, which exhibit a power-law degree distribution. Most of the work done on scale-free networks concerns off-lattice systems (graphs) where the Euclidean distance between nodes is irrelevant. However, real-life networks are often embedded in Euclidean space. For example, the Internet is embedded in a two-dimensional network of routers, and neuronal networks are embedded in a three-dimensional brain. Indeed, with the Internet, indications of the relevance of embedding space are given in [YJB02], where it is shown that the number of links (cables) of length r decreases as r^{-1} (see also Section 3.2.2).

In this section, we describe several methods for embedding scale-free networks on lattices. To be precise, we present constructions of ensembles of scale-free networks having a spatial structure that differs from the mean field ensemble presented earlier. It is plausible that when the embedding dimension is above the upper critical dimension, d_c, discussed in Section 10.4.5, these networks obey the mean-field results, i.e., the lattice geometry becomes irrelevant.

We will focus on the "lattice quota" model, presented in [RCbH02] and study some of its properties. As a guiding principle for the construction of embedded networks in general, we impose the natural restriction that the total geometrical length of links in the system be minimal. Several other models have been suggested, for instance, in

[MS02, WSS02b]. Some of these models, such as [MS02], place the nodes not on a lattice, but on randomly selected points in Euclidean space. Other models, such as [WSS02b], allow deviations from the degree distribution in the low-degree regime. The main behavior of these models is very similar in many aspects, and therefore many of the ideas presented here apply to most of them.

4.6.2 Defining the model

The "lattice quota" model is defined as follows [RCbH02]. To each node of a d-dimensional lattice, of size R, and with periodic boundary conditions, we assign a random connectivity k taken from the scale-free distribution

$$P(k) = Ck^{-\gamma}, \qquad m < k < K, \tag{4.14}$$

where the normalization constant is $C \approx (\gamma - 1)m^{\gamma-1}$ (for large K) [SK85]. We then select a node at random and connect it to its *closest neighbors* until its (previously assigned) connectivity k is realized, or until all nodes up to a distance

$$r(k) = Ak^{1/d} \tag{4.15}$$

have been explored. This ensures that the total length of the links will be minimal to a good approximation. Note that links to some of the neighboring nodes might prove impossible, if the connectivity quota of the target node is already consumed. This process is repeated for all nodes of the lattice. We show below that by following this method, networks with $\gamma > 2$ can be successfully embedded up to a (Euclidean) distance ξ, which can be made as large as desired upon increasing the external parameter A.

Suppose that one attempts to embed a scale-free network by the above recipe, in an *infinite* lattice with size, $R \to \infty$. Nodes with a degree larger than a certain cutoff $k_c(A)$ cannot be realized, because the surrounding nodes become saturated. Consider the number of links $n(r)$ entering a generic node from a surrounding neighborhood of radius r. Nodes at distance r' are linked to the origin with probability $P(k' > (r'/A)^d)$:

$$P\left(k' > \left(\frac{r'}{A}\right)^d\right) = C \int_{(\frac{r'}{A})^d} k^{-\gamma}\,\mathrm{d}k \sim \begin{cases} 1 & r' < A \\ (\frac{r'}{A})^{d(1-\gamma)} & r' > A. \end{cases} \tag{4.16}$$

Hence

$$n(r) \sim \int_0^r \mathrm{d}r' r'^{d-1} P\left(k' > \left(\frac{r'}{A}\right)^d\right)$$

$$\sim \frac{\gamma - 1}{d(\gamma - 2)} A^d - \frac{A^{d(\gamma-1)}}{d(\gamma - 2)} r^{d(2-\gamma)}. \qquad (4.17)$$

The cutoff connectivity is then

$$k_c(A) = \lim_{r \to \infty} n(r) \sim \frac{1}{\gamma - 2} A^d. \qquad (4.18)$$

The cutoff connectivity implies a cutoff length

$$\xi = r(k_c) \sim (\gamma - 2)^{-1/d} A^2. \qquad (4.19)$$

The embedded network is *scale-free* up to distances $r < \xi$, and repeats itself (statistically) for $r > \xi$, similar to the infinite percolation cluster above criticality. The infinite cluster in percolation is *fractal* up to the correlation length ξ and is repeated thereafter [BH96, bH00, SA94].

When the lattice is finite, $R < \infty$, the number of nodes is finite, $N \sim R^d$, which imposes a maximum connectivity [CEbH00, DMS01c]

$$K \sim m N^{1/(\gamma-1)} \sim R^{d/(\gamma-1)}. \qquad (4.20)$$

This implies a finite-size cutoff length

$$r_{\max} = r(K) \sim A R^{1/(\gamma-1)}. \qquad (4.21)$$

The interplay between the three length-scales, R, ξ, and $r_{\max}$ determines the nature of the network. If the lattice is finite, then the maximal connectivity is $k_{\max} = K$ only if $r_{\max} < \xi$. Otherwise ($r_{\max} > \xi$), the lattice repeats itself at length-scales larger than ξ. As long as $\min(r_{\max}, \xi) \ll R$, the finite size of the lattice imposes no serious restrictions. Otherwise ($\min(r_{\max}, \xi) > R$), finite-size effects become important. We emphasize that in all cases the degree distribution (up to the cutoff) is scale free.

In Figure 4.4(a) we show typical networks that result from this embedding method, for $\gamma = 2.5$ and 5 in two-dimensional lattices. For larger γ values, the network more closely resembles the embedding lattice, because long links become rare.[7] In Figure 4.4(b) we show the same networks as in part (a) where successive chemical shells are depicted in different shades. Chemical shell l consists of all nodes at a minimal distance (a minimal number of hops) l from a given node. For this choice of parameters, $\gamma = 5$ happens to fall in the region of $\xi > r_{\max}$, whereas for $\gamma = 2.5$,

[7] Here we choose $m = 2d$ so that in the limit $\gamma \to \infty$ the network is identical with the embedding lattice. Clearly, this choice is not mandatory.

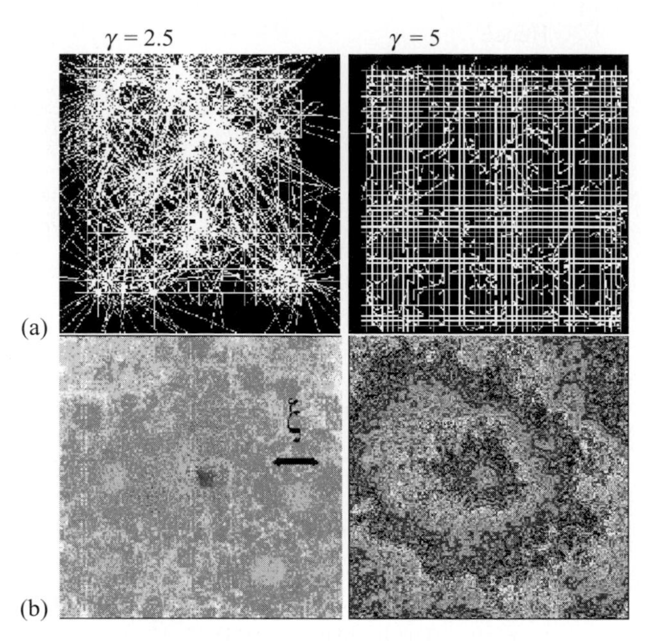

Figure 4.4 Spatial structure of the lattice quota network. (a) The typical map of links for a system of 50×50 nodes generated from degree distributions with $\gamma = 2.5$ (left) and $\gamma = 5$ (right). (b) Shells of nodes equidistant to the central one in a lattice of 300×300 nodes. Note that for $\gamma = 5$ (right), shells are concentric and continuous fractals, but for $\gamma = 2.5$ (left), shells are broken. After [RCbH02].

$\xi < r_{\mathrm{max}}$. In the latter case, we can visually observe (Figure 4.4(b), $\gamma = 2.5$) the (statistical) repetition of the network beyond the length-scale ξ.

The degree distribution resulting from this embedding method is illustrated in Figure 4.5. In Figure 4.5(a), $\xi < r_{\mathrm{max}}$ and the distribution terminates at the cutoff k_{c}. The scale-free distribution is altered slightly, for $k < k_{\mathrm{c}}$, due to saturation effects, but the overall trend is highly consistent with the original power law. The scaling in the inset confirms that $k_{\mathrm{c}} \sim A^d$. In Figure 4.5(b), $\xi > r_{\mathrm{max}}$ and the cutoff K in the distribution results from the finite number of nodes in the system. The scaling in the inset in Figure 4.5(b) confirms the known relation $K \sim m R^{d/(\gamma-1)}$ [CEbH00, DMS01c]. The different regimes are summarized in Figure 4.6.

4.6.3 Additional methods

We described a method for embedding scale-free networks in Euclidean lattices. The method is based on a natural principle of minimizing the total Euclidean length of links in the system. This principle enables us to embed the scale-free networks in Euclidean space without additional external exponents. Manna and Sen [MS02] and Xulvi-Brunet and Sokolov [WSS02b] independently suggested a different embedding

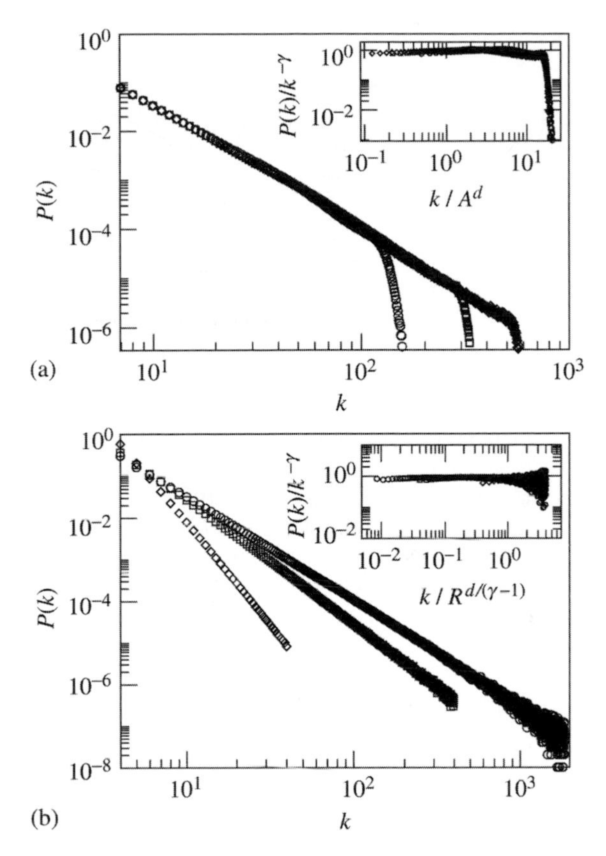

Figure 4.5

(a) The resulting degree distribution obtained from simulations performed on two-dimensional systems of size $R = 400$, $\gamma = 2.5$ and for several values of A: (circles) $A = 2$, (squares) $A = 3$ and (diamonds) $A = 4$; they all end at a cutoff $k_c(A)$ for this case where $r_{\max} > \xi$. In the inset we show a scaling collapse using the same data. The threshold takes place at $k_c \sim A^{d/(\gamma-2)}$ and confirms the validity of our theoretical estimations. (b) The power-law distribution of node degree in the network is shown for $R = 100$, $A = 10$ and for different values of γ: $\gamma = 2.5$ (circles), 3.0 (squares), and 5.0 (diamonds). Note that in all cases, the distribution achieves its (natural) cutoff K. In the inset we show the corresponding collapse supporting $K \sim R^{d/(\gamma-1)}$. In this case, $r_{\max} < \xi$. After [RCbH02].

method in Euclidean space that includes an external exponent. In Chapter 8 we will discuss the behavior of distances and the fractal dimensions of embedded scale-free networks.

4.6.4 Random geometrical graphs, continuum percolation, and ad hoc networks

In recent years the importance of wireless communication networks has increased dramatically. The usage of cellular phone and communication devices, wireless

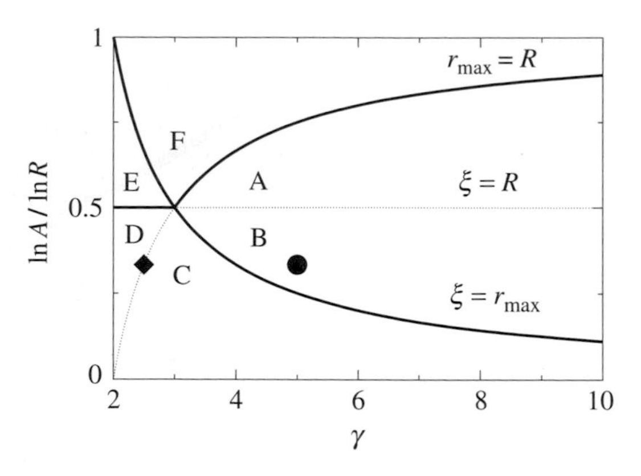

Figure 4.6 Diagram showing the six regions of different behaviors for the lattice quota network: for region A, $r_{max} < R < \xi$; B, $r_{max} < \xi < R$; C, $\xi < r_{max} < R$; D, $\xi < R < r_{max}$; E, $R < \xi < r_{max}$; F, $R < r_{max} < \xi$. The diagram can be mapped into only four regions where the cutoff k_c and the size effect K are expected. A and B, no cutoff and no size effect; C and D, cutoff and no size effect; E, cutoff and size effect; F, no cutoff but size effect. The two symbols indicate the parameters corresponding to Figure 4.4(b), (full diamond) $\gamma = 2.5$ and (full circle) $\gamma = 5$. After [RCbH02].

Ethernet, and Bluetooth equipment is becoming as widespread as regular, wired networks, and they have become a critical part of many people's daily lives. Most of these networks are not interesting in terms of their topology, and they usually depend on direct all-to-all communication or, as in cellular networks, direct communication in communication cells, where the cells themselves are connected using wired technology.

A special type of wireless network, which has received much attention is ad hoc networks. These networks are composed of units placed in geographic space at distances prohibiting direct communication between all nodes (either due to limited power, topography, or to prevent interference between units). The network, therefore, is based on each node functioning as a router, i.e., passing messages between other nodes in the network.

When all nodes transmit with the same power, it is usually assumed that the range of their transmission is similar. To model the connectivity of such a network, one can consider a model of N nodes placed randomly in a plane. To model the transmission between nodes, one must represent each node by a disk. Each disk has a radius of half the transmission range of the node, and therefore if two disks intersect, then the nodes are within the transmission range of each other. This model is identical to the model of continuum percolation [BH94, SA94], which has been studied by physicists for several decades.

Continuum percolation, as other percolation phenomena, is known to consist of a threshold density. That is, when the density of disks exceeds a certain threshold, a

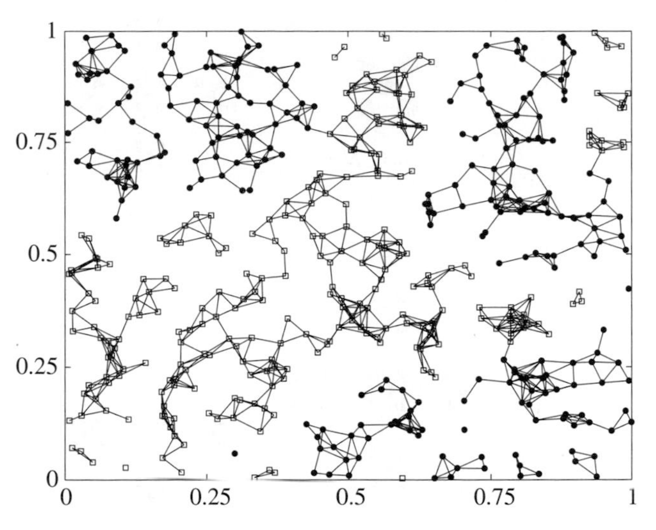

Figure 4.7 A two-dimensional random geometric graph with $N = 500$ nodes and average degree $\langle k \rangle = 5$. After [DC02].

spanning cluster appears, consisting of a finite fraction of the nodes that can exchange messages between them. No exact analytical derivation of the percolation threshold is known. However, the threshold is known numerically from simulations. Assuming that unit diameter disks are placed randomly in the plane with density ρ, the critical density is $\rho_c \approx 0.676$ [BH94, SA94].

The network above has been presented as a geometrical object. However, it can also be represented by a graph. Consider the following construction: place nodes randomly in the plane and then connect each pair of nodes only if the distance between them is at most one unit. This graph is usually referred to as the **unit disk graph** or the **random geometric graph** (see Figure 4.7 for illustration). Using the above criterion, we can calculate the average degree at the critical threshold, resulting in $\langle k \rangle_c = 4.52$ (for further information and the results for higher dimensions, see [DC02]).

To simplify algorithms running in ad hoc networks, sometimes it is not desirable to use all edges of the unit disk graph, since cliques are abundant in this graph. To simplify the graph without affecting its connectivity, we can use some known constructions that allow the dilution of edges. One well-known construction is the **Gabriel graphs**, where an edge between two nodes i and j, located at points R_i and R_j, exists only if the disk whose diameter is the line connecting the nodes $\overline{R_i R_j}$ contains no other nodes. It can be proven that this graph has exactly the same clusters as the full unit disk graph. However, it is not as dense as the original unit disk graph. To learn about more properties of the Gabriel graph and for algorithms for ad hoc networks, see, for example, [KWZZ03, KZ03].

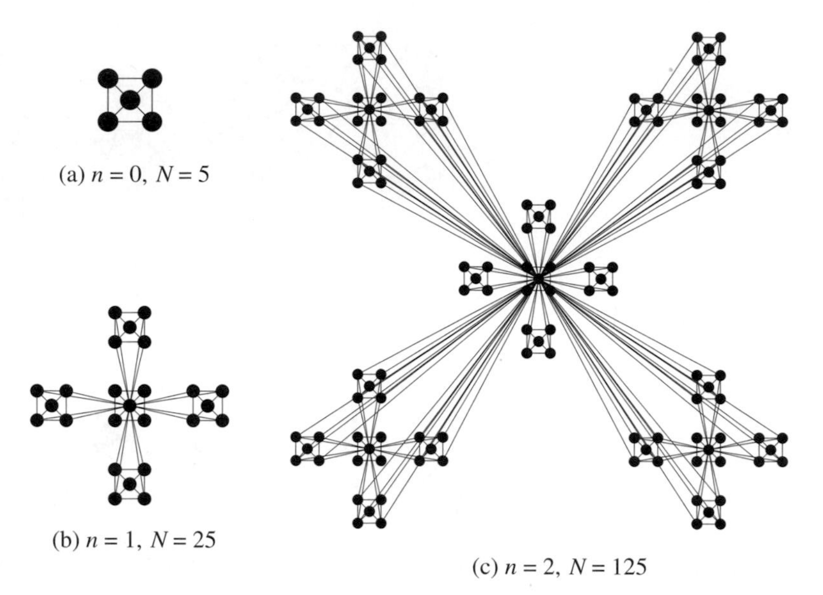

Figure 4.8 An illustration of the construction of a hierarchical network. Starting from a cluster of $N = 5$ nodes (a) at generation $n = 0$, the cluster is then duplicated four more times, and (b) each of the new duplicates is connected to the central node of the original cluster. At this generation, $n = 1$, the number of nodes is $N = 25$. The process is then repeated another time (c) for generation $n = 2$, leading to a network with $N = 125$ nodes. The process can be continued until the desired generation or size is reached. This network has a power-law degree distribution with exponent $\gamma = 1 + \ln 5/\ln 4 \approx 2.16$. After [RB03].

4.7 Hierarchical and fractal networks

Other than the degree distribution, geography, correlations and clustering, many other properties of a network can be considered. These properties lead to other classes of networks, many of which present rich and interesting behavior. One of the most interesting properties is the behavior of networks at different scales. Between the single node and full network levels, there may be many orders of magnitude. At each of these levels, the network may present a different behavior. The properties of clustering and degree-degree correlations are pronounced only at the single node or the several node level, whereas other properties, such as geographical embedding, are pronounced at all levels.

One class of networks that present interesting behavior at all orders of magnitude is the class of hierarchical networks. This class contains scale-free networks that are built by a recursive rule. This rule may be deterministic or stochastic. Applying this rule repetitively produces a network yielding a similar structure at several different orders of magnitude. Many such models have been proposed. Several examples are presented in [DGM02a, JKK02, RB03, Rb04]. Several classes of networks in nature, such as many biological networks [RSM+02] have been shown to present hierarchical structures. Figure 4.8 presents an example of a hierarchical scale-free network construction.

Fractal networks are a special case of networks with structures at every order of magnitude. They behave similarly at different orders of magnitude (at least statistically). Several networks in nature in fact show fractal features. A closer look at the properties of these networks will be presented in Chapter 7.

Spatially embedded networks present an important field of research, as many real-world networks are spatially embedded. In most of these networks, long-range links incur costs, and therefore there is a correlation between the geographical locations of neighboring nodes. For some models of spatially embedded networks see also [HBP03].

Exercises

4.1 Using the methods described in Section 4.2, find the average distance between nodes for a small-world network in a general dimension with a general k.

4.2 A bipartite network is a network with two sets of nodes, where nodes of each set are only connected to nodes from the other set. These nodes can represent two types of objects. For instance, the network of scientific collaboration can be represented by nodes representing scientists and nodes representing papers. Each scientist's node is connected to the nodes of the papers the scientist coauthored. For more information on bipartite networks, see, for example, [NSW01].

Consider a random bipartite network having degree distributions $P_a(k)$ and $P_b(k)$ for nodes of type a and b, respectively.

(a) What relation must hold between $P_a(k)$ and $P_b(k)$? (*Hint*: Every link must connect one node of each type).

(b) Devise a model similar to the configuration model (Section 4.3) for random bipartite graphs with given degree distributions (assuming they fulfill the condition in (a)).

4.3 A network can be constructed from a single set, a, of nodes of a bipartite network by connecting nodes of type a if they share a neighbor (from set b).

What is the degree distribution obtained by applying this process on a random bipartite network with degree distributions $P_a(k)$ and $P_b(k)$?

4.4 For degree-degree correlated networks, what equation should the probability function $P(k_2|k_1)$ satisfy? (See Section 4.4.)

4.5 Consider the following hierarchical construction: start from a pair of nodes connected by a link. At every generation add to each node, i, having degree k_i, k_i new links, each leading to a new node of degree 1.
(a) What is the number of nodes after n generations?
(b) What is the limiting degree distribution?

4.6 Using geometric considerations, calculate the probability of a node in a random geometric graph with density ρ to have k neighbors. That is, find the degree distribution of a random geometric graph.

4.7 Use geometric considerations to calculate the clustering of a random geometric graph with density ρ (see [DC02]).

5 Growing network models: the Barabási–Albert model and its variants

Many properties of real-world networks have been considered so far: the degree distribution, clustering, diameter, etc. However, the treatment of these properties has been rather ad hoc, i.e., each new property found has been incorporated into a simple static model. This approach has several disadvantages.

- We cannot be sure that the ensemble we use is the correct one for the problem. Usually, we try to take the largest or simplest ensemble having the desired properties. However, this may not apply to real-world networks.
- Real-world networks are usually dynamical, whether over a short time span, such as the Internet and WWW, or over a long (evolutional) time span. The classes of networks discussed so far are static, and actually are distinct for each N. No dynamical evolution of the network is incorporated.
- The equilibrium models provide no indication as to why the network received its observed properties. So far, we have treated these properties as given.

In this chapter, we attempt to address these problems by presenting models for network creation and evolution. These growth models reproduce several observed properties of real-world networks using simple ideas. Most of these models are based on the simple Barabási–Albert model that is presented next.

5.1 The Barabási–Albert model

The Barabási–Albert model [BA99] is based on two simple assumptions regarding network evolution.

(i) **Growth:** new nodes are being added to the network, where each new node is connected to m existing nodes.
(ii) **Preferential attachment:** this is the heart of the model. Each new node is connected to existing nodes with a probability proportional to its existing degree.

In a more rigorous manner, consider a network evolving in time, t, where at each time unit a new node is added to the network and connected to m existing nodes, where the probability of connecting to an existing node, i, $\Pi(i)$ is given by

$$\Pi(i) = \frac{k_i}{\sum_j k_j}. \tag{5.1}$$

The initial kernel at time $t = 0$ is usually assumed to be connected, but the details of its structure have only a small effect on the final result. Examples of possible selections are a single node or a clique of $m + 1$ nodes.

5.1.1 Analysis

There are several methods for analyzing the results of the Barabási–Albert model. The simplest method is a mean-field analysis originally described in [BA99]. In this method the degree of each node is treated as a single-valued function of its creation time, s, and the current time, t. A dynamical equation is written for the degree of each node:

$$\frac{\partial k_s}{\partial t} = m \frac{k_s}{\sum_{j=1}^{N-1} k_j}, \tag{5.2}$$

where N is the (current) number of nodes. Since at each time unit a new node is formed and connected to m other nodes, $2m$ links are added at each time unit. Therefore, the total number of links is $\sum_{j=1}^{N-1} k_j = 2mt$ (assuming the network started at time $t = 0$ with m self loops). This leads to

$$\frac{\partial k_s}{\partial t} = \frac{k_s}{2t}, \tag{5.3}$$

with the initial condition $k_s(t = s) = m$ (except for the first node, where the degree begins with $2m$), the solution being $k_s = m(t/s)^{1/2}$. Using the transformation between the variables s and k (see Appendix A), and noting that since the rate of new node creation is 1 (i.e., $p(s) = 1/t$), one obtains

$$p(k) = p(s)\left|\frac{\mathrm{d}s}{\mathrm{d}k}\right| = \frac{2m^2}{k^3}, \quad k < t\,. \tag{5.4}$$

Thus, this model leads to a power-law degree distribution, with $\gamma = 3$.

Other methods for analyzing this model using tools from statistical physics include the master equation approach [DMS00] and the rate equation approach [KRL00]. We will discuss the master equation approach here. The interested reader is referred to [KRL00] for the rate equation approach.

In the master equation approach, an equation is written for the probability that a node, s, has degree k at time t. This approach is more accurate than the mean-field approach, since it allows the degree of each node to be selected from some distribution, rather than being a deterministic function of the creation and current times. This approach does, however, assume that the node degrees are independent random variables, and are not correlated between them. The master equation is:

$$p(k,s,t+1) = \frac{k-1}{2t}p(k-1,s,t) + \left(1 - \frac{k}{2t}\right)p(k,s,t), \tag{5.5}$$

where $p(k, s, t)$ is the probability of a node born at time s to be of degree k at time t. The two terms represent the probability of a node having degree $k-1$ gaining a new link and of a node having degree k gaining a new link, respectively. It is assumed that at time $t = 1$, a pair of nodes, $s = 0, 1$, connected by a link is present. Therefore, the initial condition is $p(k, s = 0, 1, t = 1) = \delta_{k,1}$ and the boundary condition due to the formation of new nodes is $p(k, t, t) = \delta_{k,1}$. Equation (5.5) may be rewritten in the form

$$2t[p(k,s,t+1) - p(k,s,t)] = (k-1)p(k-1,s,t) - kp(k,s,t)\,. \tag{5.6}$$

Passing to the continuous limit in t and k, we obtain

$$2t\frac{\partial p(k,s,t)}{\partial t} + \frac{\partial[kp(k,s,t)]}{\partial k} = 0 \tag{5.7}$$

and

$$\frac{\partial[kp(k,s,t)]}{\partial \ln\sqrt{t}} + \frac{\partial[kp(k,s,t)]}{\partial \ln k} = 0\,. \tag{5.8}$$

The solution of Eq. (5.8) is $kp(k, s, t) = \delta(\ln k - \ln\sqrt{t/s} + \text{constant})$. The boundary condition is fulfilled if it is of the following form

$$p(k,s,t) = \delta(k - \sqrt{t/s})\,. \tag{5.9}$$

Therefore, we see that the transition to the continuous limit in the master equation leads to the δ-function form of the degree distribution of individual nodes.

The main quantity of interest is the total degree distribution of the entire network,

$$P(k,t) = \frac{1}{t+1}\sum_{s=0}^{t} p(k,s,t)\,. \tag{5.10}$$

In the continuous approximation, the stationary degree distribution is of the form

$$P(k) = P(k, t\to\infty) = \lim_{t\to\infty}\frac{1}{t}\int_0^t \mathrm{d}s\; p(k,s,t)\,. \tag{5.11}$$

Inserting the obtained expression for $p(k, s, t)$, Eq. (5.9), into Eq. (5.11), one obtains the continuous approximation result for this model [BA99, BAJ00],

$$P(k) = 2/k^3\,. \tag{5.12}$$

The rate equation approach focuses on the number of nodes of degree k, $N_k(t)$, rather than on the dependence of the degree on the birth and current times. In [KRL00] a non-linear growing model (see Section 5.2.1) is studied using this method, giving the Barabási–Albert model as a special case.

5.2 Variants of the Barabási–Albert model

The Barabási–Albert model is characterized by producing a power-law degree distribution using a very simple model. However, it should be viewed as a simplification of reality, since it does not take into account many properties of evolving real-world networks. Some of these properties are correlations between different nodes, stemming from some sort of internal properties, the changing nature of nodes over time (aging), and non-linear preferential attachment effects. Furthermore, the phenomenon of preferential attachment assumes use of linear preference in the existing degree. Although this seems very plausible, an explanation is still necessary as to why such linear preferential attachment is expected.[1]

5.2.1 Non-linear growing models

Some studies of the effect of deviating from the linear attachment rule can be found in [DMS01c, KR01, KRL00]. In [DMS01c] the effect of a shifted linear preference rule,

$$\Pi(i) = \frac{k_i + A}{\sum_j (k_j + A)}, \tag{5.13}$$

for some constant A is studied.

The solution of the shifted linear model leads to a degree distribution[2] $P(k) \sim k^{-\gamma}$, with

$$\gamma = 3 + \frac{A}{m}, \quad A > -m, \tag{5.14}$$

where the restriction $A > -m$ guarantees that a new node can connect to every existing node with a positive probability. As can be seen from Eq. (5.14), this model leads to a power-law degree distribution with $2 < \gamma < \infty$. Naturally, γ cannot obtain

[1] For a study showing that linear preferential attachment actually exists in evolution, see [EL03].

[2] In fact, in [DMS01c] a more general model is investigated with the possibility of other, “virtual,” links.

a value of 2 or less since the total number of links is proportional to the number of nodes. However, other than that restriction, this model leads to a tunable power-law degree distribution. This indicates that the model is not stable in the produced degree distribution, i.e., any small deviation from the perfectly linear preference leads to a different behavior of the tail of the degree distribution.

The case of a mixed preferential attachment model, where at every stage some of the links are connected according to a preferential attachment rule and some are connected randomly, is very similar to the shifted linear preferential attachment, since in both cases, an intrinsic attractivity competes with the preferential attachment.

Another possible extension of the preferential attachment model is a non-linear preferential attachment law [KR01], where the simplest form is a non-linear power law

$$\Pi(i) = \frac{k_i^\alpha}{\sum_j k_j^\alpha} . \tag{5.15}$$

In the case where $\alpha \neq 1$, the growth model no longer leads to a power-law degree distribution. In fact, several regimes exist; in each a different behavior is observed.

- In the sublinear regime, where $\alpha < 1$, a power-law degree distribution with an exponential cutoff occurs. This implies that the sublinear preferential attachment is not strong enough to produce a pure power-law degree distribution.
- The linear case $\alpha = 1$ leads to a tunable exponent degree distribution with $2 < \gamma < \infty$, as discussed above. The pure linear case, in which the additive constant $A = 0$ is a special point in this regime, corresponds to the original Barabási–Albert model with $\gamma = 3$.
- In the super-linear regime, $\alpha > 1$, a "condensation" to a gel-like state occurs. This regime is the extreme case of the "winner takes all" principle, where a single node is connected to almost all nodes in the network. Again, the power law disappears, and a degenerate form of a degree distribution appears, where a single node has degree of order $\mathcal{O}(N)$. See [KRL00] for details.

Thus, it can be seen that the models that produce a power-law degree distribution represent a narrow regime around the linear preferential attachment line. The original Barabási–Albert model is a single point on this line. Therefore, the existence of so many power-law networks, or nearly power-law networks, seems to indicate some natural attraction towards the linear preferential attachment regime (see [EL03, JNB03, PVV01] for some direct observation). An example of a mechanism attempting to account for the linear preferential attachment is described in Section 5.2.3.

5.2.2 Rewiring and aging

The assumption that nodes and the links between them remain static throughout the life of the network seems inappropriate for most real-world networks. Routers and whole networks in the Internet have been added, removed, and replaced throughout the years. Links are added and removed between routers. The WWW is highly dynamic, with the contents of pages and websites changing daily, and sometimes hourly. Webpages may disappear or change dramatically, and links are easily removed and added even for long-existing pages. Similar changes appear in transportation and man-made networks. Links and nodes in biological networks also change in time, mainly through the slow process of evolution.

To account for all these changes, several models have been proposed, containing effects such as rewiring, where existing links may be deleted or replaced with other links, and also links between existing nodes may appear at any time (whereas, in the Barabási–Albert model the links are only added when a new node appears). Another effect that may be taken into account is aging – where the attractiveness of a node not only depends on its current degree, but also decays with time. This models the decrease in a node's appeal as it becomes older, compared with the new competition. For a survey of results for many of the above models, see, for example, [DM02, DM03].

5.2.3 Copying models

A possible mechanism for the preferential attachment observed in many networks is suggested in [KR05]. Consider a citation network, where each new node, representing an article, cites some of the previous work. Whenever authors cite a previous article, assumed to be chosen at random, they may also, with some probability, cite one of the references of the cited article. The probability of selecting an article for citation using the copying mechanism is approximately proportional to the number of articles citing it. This induces a natural form of linear preferential attachment, which leads to power-law degree distributions.

A variant of this model is the case where a new node copies all the links of an existing node. This is believed to be relevant for genetic networks, where genes are sometimes duplicated. Therefore, all the function and interaction patterns are duplicated with them. The genes then undergo specialization by slightly differentiating themselves somewhat from their copy. This model again leads to approximate linear preferential attachment and to power-law degree distribution.

5.3 Linearized chord diagram (LCD)

The Barabási–Albert model contains some freedom in the way it is defined. Properties such as the initial speed and the decision whether to recalculate the probability to connect to every node after each link is added (when $m > 1$) can influence the ensemble of networks formed by the model. Although, as stated above, these decisions have a slight influence on the eventual degree distribution and on the macroscopic properties of the network (such as the diameter and the robustness), they may be interesting for some applications. In an attempt to devise a rigorous comparable model, Bollobás and Riordan [BR04] suggested the linearized chord diagram (LCD) model. One interesting feature of this model is that, at least in its common presentation, it is a static model. However, it is closely related to the Barabási–Albert model, and has many similar properties.

The LCD construction starts with $2N$ nodes (or $2mN$ nodes for some m) arranged in a line. Then, a random matching of the nodes is selected, i.e., the nodes are chosen pair by pair without replacement, until all nodes are paired. Then, the line is traversed left to right (assuming the first node is located at the left) and the nodes and links between them are modified using the following rule: starting at the first node, the nodes are collected, until a node carrying a link going towards the left is reached. All the nodes, starting from the first, up until this node, are unified into one supernode, which will be a single node in the generated network. The process continues, starting at the node immediately following the last unified one until no nodes are left. For a general m, the same process is performed, unifying nodes into a supernode only after m left-pointing links are encountered. An illustration can be seen in Figure 5.1.

Since every edge connects a pair of nodes, for one of which it points to the right, and for the other it points to the left, exactly N (super)nodes are formed in the eventually formed network. To determine the degree distribution of the (super)nodes, we will use the following mean-field approximation. Take the ith node in the original line. For simplicity we define the variable $t = i/2N$. Therefore, t changes in the range $0 \leq t \leq 1$.

Since the matching on the nodes is random, the ith node has probability t of linking to a node on its left and probability $1 - t$ of connecting to a node on its right. A supernode is formed when another supernode ends (via a left-pointing link), and includes all the following right-pointing links and is finalized by the next left-pointing link. The distribution of the number of nodes (and therefore also the links) in the supernode is therefore geometric, with probability $1 - t$ of continuing and probability t of finalizing the supernode.[3] Thus, the average degree of a supernode starting at

[3] Note that t changes from node to node in the supernode. However, for large N values, this is negligible.

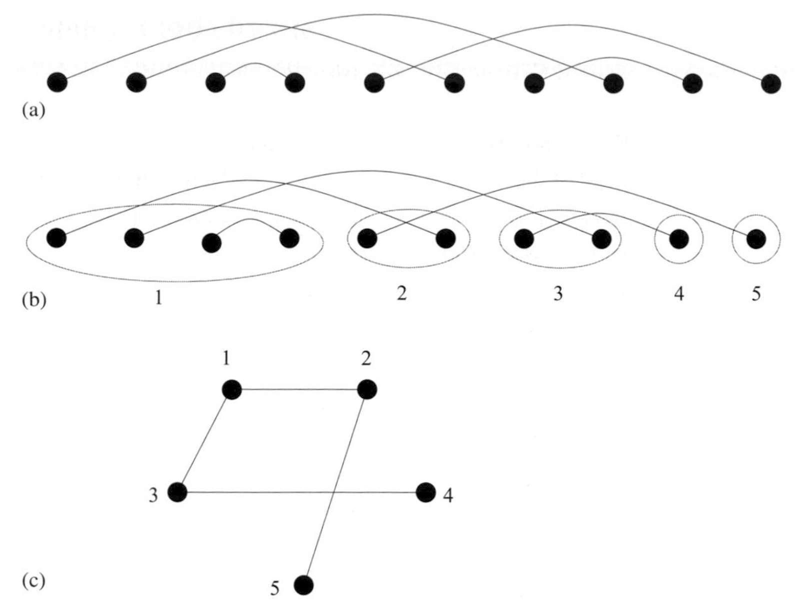

Figure 5.1 Illustration of the LCD model. (a) The linear network with the original nodes. (b) Unifying nodes into supernodes – each group ends with a link pointing leftwards. (c) The eventual network: each of the groups in (b) forms a node in the eventual network.

linear node i is

$$\sum_j (j+1)t(1-t)^j = \frac{1}{t} . \tag{5.16}$$

For simplicity, we assume that the degree of the supernode starting at node i is always $k_i = 1/t$. The probability that a node starts at node i is the supernode density near node i, which is the inverse of the number of nodes in an average supernode near i. Thus, the density of supernodes starting at $t = i/2N$ is $N(t) = 2Nt$, and the probability of a supernode starting at node i is $P(t) = N(t)/N = 2t$. The probability of forming nodes having degree k is therefore (see Appendix A for an explanation)

$$P(k) = P(t) \left| \frac{\mathrm{d}t}{\mathrm{d}k} \right| = \frac{2t}{k^2} = 2k^{-3}, \tag{5.17}$$

which is exactly the same as in the Barabási–Albert model.

The long-tailed nature of the degree distribution in the LCD model results from the bias of the probability of a link to point left or right according to its linear location. The "old" nodes (small t) tend to have a high degree, since the probability of a link starting there to point left is low. On the other hand, "new" nodes (t close to 1) tend to have few links, most of which point right, i.e., to older supernodes. For a more exact analysis of the LCD model, see [BR04].

5.4 Fitness models

One property of the Barabási–Albert model and its variants is the conservation of the hubs over time. A latecomer node will have an exceedingly small probability of obtaining the highest (or one of the highest) degrees. The maximum degree will almost certainly belong to the chronologically first node (or one of the first nodes). However, in real-world networks, some nodes may overthrow the older nodes owing to their high internal fitness. Internal fitness models have been proposed to model this property of real-world networks [BB01].

In the model presented in [BB01], it is assumed, similarly to the Barabási–Albert model, that at every time step a node is added to the network along with m links connecting this node to existing nodes. Each node, i, is also assigned a fitness parameter, η_i, chosen from a distribution $\rho(\eta)$. The probability that a newly formed node will connect via one of its links to an existing node j is proportional to both node j's fitness and its current degree:

$$\Pi_i = \frac{\eta_j k_j}{\sum_\ell \eta_\ell k_\ell} \,. \tag{5.18}$$

This model can be mapped to a Bose–Einstein gas [BB01] using the following mapping. Every node is assigned an energy level $\epsilon_i = -(1/\beta)\ln \eta_i$, where $\beta \equiv 1/T$ plays the role of the inverse temperature. The addition of a new node, i, is modeled as the introduction of a new energy level, ϵ_i, as well as the introduction of $2m$ particles, corresponding to the $2m$ ends of the links added. Each edge added corresponds to the addition of two particles to the appropriate energy levels of both its end nodes. Once these particles are added, they can no longer change state, which is different from standard thermodynamic treatment. However, at the limit of $t \to \infty$, the model converges to a thermodynamic equilibrium state.

Consider the number of particles at level ϵ_i (corresponding to a node added at time i) at time t. The rate equation for this number of particles is

$$\frac{\partial k_i(\epsilon_i, t, i)}{\partial t} = m \frac{\mathrm{e}^{-\beta\epsilon_i} k_i(\epsilon_i, t, i)}{Z_t} \,, \tag{5.19}$$

where Z_t corresponds to the partition function (or normalization factor) at time t, which is

$$Z_t = \sum_{j=1}^{t} \mathrm{e}^{-\beta\epsilon_i} k_i(\epsilon_i, t, i) \,. \tag{5.20}$$

Assume that the degrees k_i follow a power law (the consistency of this assumption will follow later)

$$k_i(\epsilon_i, t, i) = m \left(\frac{t}{i}\right)^{f(\epsilon_i)} \,, \tag{5.21}$$

for some function of the energy level, $f(\epsilon_i)$. Since η is chosen from the probability distribution $\rho(\eta)$, the energy is distributed according to $g(\epsilon) = \beta\rho(\mathrm{e}^{-\beta\epsilon})\mathrm{e}^{-\beta\epsilon}$. The average partition function can then be determined by averaging over the energy levels and the node formation times:

$$\langle Z_t \rangle = \int \mathrm{d}\epsilon g(\epsilon) \int_1^t \mathrm{d}t_0 \mathrm{e}^{-\beta\epsilon} k(\epsilon, t, t_0) \approx \frac{mt}{z}, \tag{5.22}$$

where,

$$z^{-1} = \int \mathrm{d}\epsilon g(\epsilon) \frac{\mathrm{e}^{-\beta\epsilon}}{1 - f(\epsilon)} \,. \tag{5.23}$$

A chemical potential, μ, satisfying $\mathrm{e}^{-\beta\mu}$, can now be defined. In the limit of $t \to \infty$, using Eqs. (5.22) and (5.23), we obtain

$$\mathrm{e}^{-\beta\mu} = \lim_{t\to\infty} \frac{\langle Z_t \rangle}{t} \,. \tag{5.24}$$

Using Eqs. (5.19) and (5.24), Eq. (5.21) leads to

$$f(\epsilon) = \mathrm{e}^{-\beta(\epsilon-\mu)} \,. \tag{5.25}$$

Using Eqs. (5.23) and (5.25), it follows that the chemical potential is the solution of

$$I(\beta, \mu) \equiv \int \mathrm{d}\epsilon g(\epsilon) \frac{1}{\mathrm{e}^{\beta(\epsilon-\mu)} - 1} = 1 \,. \tag{5.26}$$

In spite of the inertness of the "particles," at $t \to \infty$ the obtained expression is the same expression obtained for a Bose–Einstein gas, where $g(\epsilon)$ represents the level statistics, and $n(\epsilon) = (\mathrm{e}^{\beta(\epsilon-\mu)} - 1)^{-1}$ is the level occupation.

As in a Bose–Einstein gas, the chemical potential is always non-positive, and thus $I(\beta, \mu)$ obtains its maximum when $\mu = 0$. If $I(\beta, 0) < 1$, no solution exists for Eq. (5.26), and the continuous treatment breaks down. This leads to the well-known Bose–Einstein condensed phase. In **Bose–Einstein condensation** the continuum approximation of the level statistics breaks down, since a finite fraction of the particles in the system occupy a single level – the lowest energy level. In this case the mass conservation equation

$$2mt = \sum_{i=1}^{t} k(\epsilon_i, t, i) = mt + mtI(\beta, \mu), \tag{5.27}$$

becomes

$$2mt = mt + mtI(\beta, \mu) + n_0(\beta), \tag{5.28}$$

where $n_0(\beta)$ is the fraction of particles occupying the lowest energy level. Thus, $n_0(\beta)$ can be deduced from the equation

$$\frac{n_0(\beta)}{mt} = 1 - I(\beta, 0) \,. \tag{5.29}$$

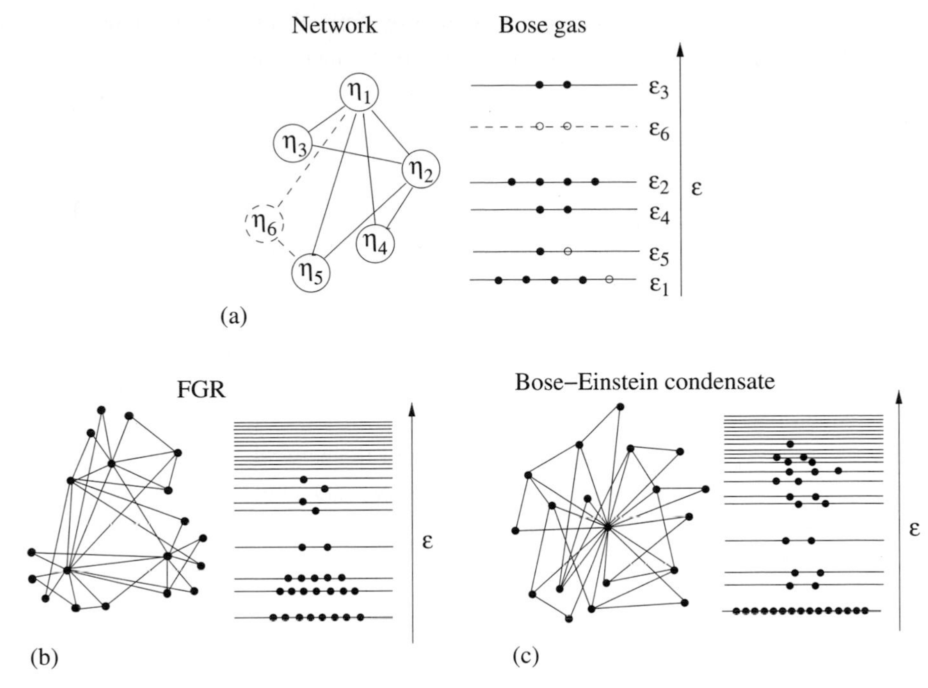

Figure 5.2 (a) The mapping of the fitness model to a Bose gas. (b) The fit-get-rich (FGR) phase, where the fittest nodes become the hubs. (c) The condensed phase, where a single node gets a finite fraction of all links in the network. After [BB01].

Three possible phases exist for this system.

(i) **The scale-free phase:** when all nodes have the same fitness, $\rho(\eta) = \delta(\eta - 1)$, or $g(\epsilon) = \delta(\epsilon)$. In this case the fitness has no influence on the model, and only the preferential attachment is pronounced. Therefore, the treatment is exactly the same as in Section 5.1, and the degree distribution obtained there, $P(k) \sim k^{-3}$, is obtained.

(ii) **The fit-get-rich phase:** at the thermodynamic limit, $t \to \infty$, the high fitness nodes become the hubs. The degree distribution can be deduced using Eq. (5.21). The degree of the highest fitness nodes scales as $k(\epsilon_{\min})/(mt) \sim t^{f(\epsilon_{\min})-1}$. The tail of the degree distribution is determined by the behavior of $\rho(\eta)$ at $\eta \to \infty$.

(iii) **The condensed phase:** when $I(\beta, 0) < 1$ a finite fraction of the links connect to the single node with highest fitness. Indeed, consider a network with level statistics $g(\epsilon) = C\epsilon^{\theta}$, $0 < \epsilon < \epsilon_{\max}$ where C is the normalization factor, $C = (\theta + 1)/\epsilon_{\max}^{\theta+1}$. The integral in Eq. (5.26) then becomes

$$I(\beta, 0) = \frac{\theta + 1}{(\beta\epsilon_{\max})^{\theta+1}} \int_{\beta\epsilon_{\min}(t)}^{\beta\epsilon_{\max}} \mathrm{d}x \, \frac{x^{\theta}}{\mathrm{e}^{x} - 1} \,. \tag{5.30}$$

Extending the integration limits from zero to infinity and comparing the integral to one, it can be seen that a lower bound for the transition temperature, $T_{BE} = 1/\beta_{BE}$ is

$$T_{BE} > \epsilon_{max}[\zeta(\theta+1)\Gamma(\theta+2)]^{-1/(\theta+1)}, \tag{5.31}$$

where $\zeta(x)$ is the Riemann zeta function of x. Thus, the transition to the Bose–Einstein condensate phase occurs at some finite temperature.

A schematic illustration of the mapping of the fitness model to the Bose–Einstein gas, and of the different phases can be seen in Figure 5.2. Note that the temperature in this model is not a feature of the fitness distribution or any other topological parameter. It is merely a parameter used for the mapping to the gas model, and therefore the actual topological structure of the network does not depend on the choice of T, but rather on $\rho(\eta)$ alone. Therefore, each choice of T would eventually lead to the same topological structure.

Exercises

5.1 Solve the model of shifted linear preferential attachment (Section 5.2) and find the degree distribution of the formed networks (see [DMS01c] for a solution).

5.2 A network grows by adding nodes, where every new node is connected randomly to an existing node, and then with probability p to a neighbor of its neighbor. Using a mean-field approach, find the degree distribution and clustering coefficient of the formed network.

5.3 In a network growth model, at every time step a new node is added, and it is connected to m nodes, where each link is connected according to the linear preferential attachment rule (Eq. (5.1)) with probability p, and to a random existing node with probability $1-p$. Map this model into a shifted linear preferential attachment model (Section 5.2) and find the degree distribution (see [DMS01c] for a solution).

5.4 Write the rate equations for N_k, the number of nodes of degree k in the Barabási–Albert model, and solve them to obtain the degree distribution of the network (see [KRL00] for a solution).

5.5 Solve the model of preferential attachment with $\Pi(j) \propto k_j^{\alpha}$ (Section 5.2) and find the degree distribution of the formed networks for the different regimes of α (see [KRL00] for a solution).

PART II

STRUCTURE AND ROBUSTNESS OF COMPLEX NETWORKS

6 Distances in scale-free networks: the ultra small world

6.1 Introduction

It is well known [AJB99, Bol85, CL01, BNKM01] that random networks, such as Erdős–Rényi networks [ER59, ER60] as well as partially random networks, such as small-world networks [WS98], have a very small average distance (or diameter) between nodes, which scales as $d \sim \ln N$, where N is the number of nodes. Since the diameter is small even for large N values, it is common to refer to such networks as "small-world" networks. Many natural and man-made networks have been shown to possess a scale-free degree distribution, including the Internet [FFF99], WWW [AJB99, BKM+00], metabolic [JTA+00] and cellular networks [JMBO01], as well as trust cooperation networks [GGA+02] and email networks [EMB02]. For more details, see Chapter 3.

The question of the diameter of such networks is fundamental. It is relevant in many fields regarding communication and computer networks, such as routing [GKK01b], searching [ALPH01], and transport of information [GKK01b]. All these processes become more efficient when the diameter is smaller (see, e.g., [LCH+06]). It might also be relevant to subjects such as the efficiency of chemical and biochemical processes and the spreading of viruses, rumors, etc. in cellular, social, and computer networks. In physics, the scaling of the diameter with the network size is related to the physical concept of the dimensionality of the system, and is highly relevant to phenomena such as diffusion, conduction, and transport in general. The anomalous scaling of the diameter in these networks is expected to lead to anomalies in diffusion and transport phenomena on these networks. In this chapter, we investigate the diameter of scale-free random networks and show that it is significantly smaller than the diameter of regular random networks. We show that scale-free networks with $2 < \gamma < 3$ have diameter $d \sim \ln \ln N$ and thus can be considered as "ultra small-world" networks [CH03].

The psychologist Stanley Milgram [Mil67] was the first to study the diameter of a real social network. In the 1960s he conducted research on the social distance between individuals in the USA by sending letters to random individuals in Nebraska, asking each of them to try to forward the letter to another random individual in Boston by

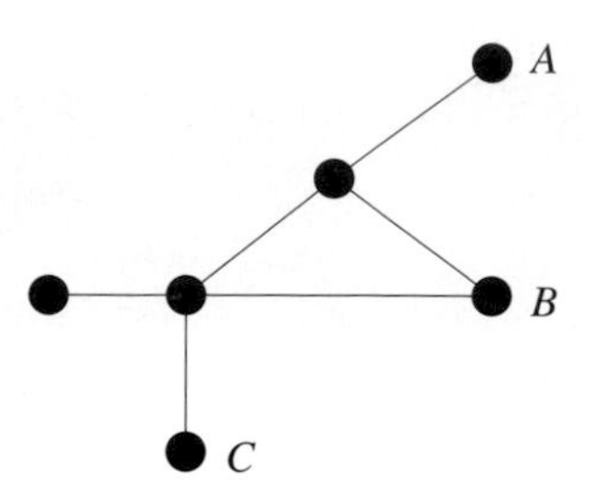

Figure 6.1 An illustration of distances in graphs. In this example, $d(A, B) = 2$, $d(A, C) = 3$, $d(B, C) = 2$. The sum of all distances in this graph is 52 and there are 30 (ordered) pairs of nodes. Therefore, the diameter (average distance) is $52/30 \approx 1.733$. The maximum distance is 3.

the shortest path they could come up with, and only to transfer the letters through people they were acquainted with on a first-name basis. Of those letters that reached their destination, the average distance traveled was about 6 hops. This phenomenon has come to be known as the "six degrees of separation," and is the basis of many books, a play, and many other studies.

We define the average diameter of a graph as the average distance between all pairs of nodes in the graph (unlike the usual mathematical definition of the diameter as the largest distance between two nodes). Since no embedding space is defined for these networks, the distance denotes the shortest path between two nodes (i.e., the smallest number of hops (over links) needed to reach one node from the other). See Figure 6.1 for an illustration of distances in graphs. If the network is fragmented, we will only be interested in the diameter of the largest cluster (assuming there is one).

To estimate the diameter, we will study the radius of such graphs. We define the radius of a graph as the average distance of all nodes on the graph from the node with the highest degree in the network (if there is more than one such node, we will arbitrarily choose one of them). The diameter of the graph, d, is restricted to:

$$r \leq d \leq 2r, \tag{6.1}$$

where r is the radius of the graph, defined as the average distance $\langle l \rangle$ between the highest degree node (the "origin") and all other nodes.

A scale-free graph is a graph having degree distribution, i.e., the probability that a node has k links (see Eq. (2.2))

$$P(k) = ck^{-\gamma}, \quad k = m, m+1, \ldots, K, \tag{6.2}$$

where $c \approx (\gamma - 1)m^{\gamma-1}$ is a normalization factor, and m and K are the lower and upper cutoffs of the distribution, respectively. The ensemble of such graphs has been defined in [ACL00]. However, we will refer here to the ensemble of scale-free graphs with the "natural" cutoff[1] $K = mN^{1/(\gamma-1)}$ (Eq. (2.4)).

6.2 Minimal distance networks

We begin by showing that the lower bound on the diameter of any scale-free graph with $\gamma > 2$ is on the order of $\ln \ln N$. Then, we show that for random scale-free graphs with $2 < \gamma < 3$, the diameter actually scales as $\ln \ln N$. For $\gamma > 3$, the diameter scales as $d \sim \log N$, similar to ER networks. Thus, random scale-free networks can be regarded, in this aspect at least, as a generalization of ER networks.

One can clearly see that the smallest diameter for a graph with a given degree distribution is found by the following construction: start with the highest degree node, and then in each layer attach the next highest degree nodes until the layer is full. By construction, loops will occur only in the last layer. This structure is somewhat similar to a graph with assortative mixing [New02a], – since high-degree nodes tend to connect to other high-degree nodes.

In this kind of graph the number of links outgoing from the lth layer (nodes at distance l from the origin), χ_l, equals the total number of nodes with degrees between K_l, which is the highest degree of a node not reached in the lth layer, and K_{l+1}, which is the same for the $l+1$ layer (see Figure 6.2). This can be described by the following equation:

$$\chi_l = N \int_{K_{l+1}}^{K_l} P(k)\mathrm{d}k \approx m^{\gamma-1} N K_{l+1}^{1-\gamma} . \tag{6.3}$$

The number of links outgoing from the $l+1$ layer equals the total number of links in all the nodes between K_l and K_{l+1} minus one link at every node, which is used to connect to the previous layer:

$$\chi_{l+1} = N \int_{K_{l+1}}^{K_l} (k-1)P(k)\mathrm{d}k \approx \frac{\gamma-1}{\gamma-2} m^{\gamma-1} N K_{l+1}^{2-\gamma} . \tag{6.4}$$

Solving those recursion relations with the initial conditions $K_0 = N^{1/(\gamma-1)}$ and $\chi_0 = K_0$ leads to:

$$\chi_l = a^{(\gamma-1)(1-u^l)} N^{1-u^{l+1}} , \tag{6.5}$$

[1] This cutoff is chosen since this is the "natural" cutoff of the network, due to the finite size of the system, if no external limitation is imposed, see [CEbH00, DMS01a]. However, as can be seen from Eqs. (6.7) and (6.10), every cutoff above $\sqrt{N}$ will give similar results. For the deterministic graphs, every power of N will give similar results.

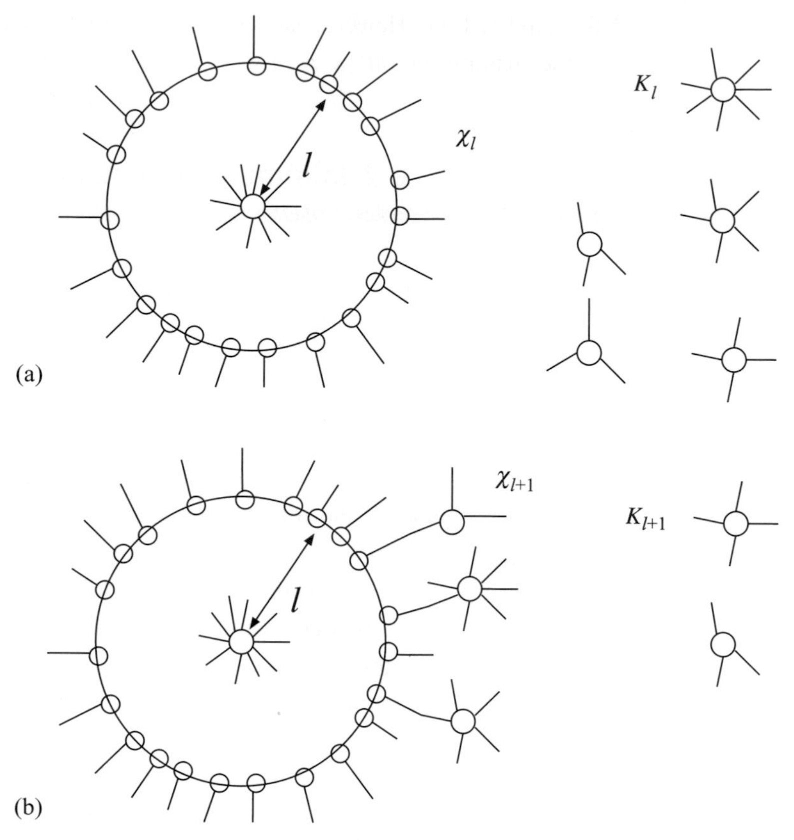

Figure 6.2 Illustration of the exposure process. The large circle denotes the exposed fraction of the giant component, whereas the small circles denote individual nodes. The nodes on the right have not been reached yet. (a) The structure after the exposure of the lth layer. (b) The structure after the exposure of the $l+1$ layer. After [CH03].

where $a = (\gamma - 1)/(\gamma - 2)m$, $u = (\gamma - 2)/(\gamma - 1)$, and

$$K_l = m(\chi_l/N)^{\frac{1}{1-\gamma}} . \tag{6.6}$$

To bound the radius, r, of the graph, we will assume that the low-degree nodes are connected randomly to the center. We choose some degree $1 \ll k^* \approx (\ln \ln N)^{1/(\gamma-1)}$. We can use Eq. (6.6) to show that if $l_1 \approx \ln \ln N / \ln(\gamma - 2)$, then $K_{l_1} < k^*$, so nodes with degrees $k \geq k^*$ would have been reached with probability approaching 1 in the first l layers.[2] On the other hand, if we start uncovering the graph from any node – provided that it belongs to the giant component – then at a distance l_2 from this node, there are at least l_2 links. The probability that none of those links will lead to a node

[2] Note that since there are $\mathcal{O}(N)$ nodes of degree 1 and 2, there are nodes of degrees that are at a longer distance from the core than $\ln \ln N$, and, in fact, the longest distances are of order $\ln N$ [CL01]. However, the majority of nodes are at distance $\mathcal{O}(\ln \ln N)$, and this is also the average distance.

of degree k^* decays as $(1 - k^* P(k^*)/\langle k \rangle)^{l_2}$. So, assuming $k^{*\gamma-1} \ll l_2 \ll \ln \ln N$, we will, with high probability, reach a node of a degree at least k^* at distance l_2 from almost every node. Since $l = l_1 + l_2$, all those nodes are at a distance of order $\ln \ln N$ from the highest degree node; this is the behavior of the radius of the graph. Thus, $\ln \ln N$ is a lower bound for the diameter of scale-free networks, and by applying this approach, one can generate scale-free networks of this smallest diameter, for any $\gamma > 2$.[3]

For $\gamma = 2$ the construction is somewhat similar to the condensate obtained in [BB01]. In this case, the highest degree node has $\mathcal{O}(N)$ neighbors; therefore, a finite fraction of the nodes will connect to this node, and all the rest can be connected at the second layer. Thus, the distance is constant and independent of N.

6.3 Random scale-free networks

In the following, we present analytical arguments showing that the behavior of $d \sim \ln \ln N$ is actually achieved in random non-correlated scale-free graphs with $2 < \gamma < 3$. Non-correlated networks are networks in which the degree of a node reached by following a link is independent of the degree of the node at the other side of that link. One can view the process of uncovering the network (which is the same as building it) by following the links one at a time. For simplicity, let us start with the node having the highest degree (which is also guaranteed to belong to the giant component), whose degree is proportional to $N^{1/(\gamma-1)}$ [CEbH00, DMS01a]. Next, we expose the layers, $l = 1, 2, 3, \ldots$, one at a time. To this end, we consider the graph as built from one large developing cluster, and nodes that have not yet been reached (they can also belong to the giant component or not belong). Molloy and Reed [MR98] considered the graph similarly.

After layer l is explored, the distribution of the unreached nodes changes (since most high-degree nodes are reached in the first layers). To take this into account, we assume that the lth layer has χ_l outgoing links. The distribution of degrees, after uncovering some of the edges, changes to $P'(k) \approx P(k)\exp(-k/K_l)$ [MR98]. In the limit of large N and large K_l values, we will assume that after exploring this layer, the highest degree of the unvisited nodes is of order K_l, where χ_l and K_l are functions of l that will be determined later.

[3] It should be noted that while this is the behavior for the average distance, the maximum distance in random scale-free graphs with $m < 3$ is larger, and scales as $\log N$. This is due to the fact that almost every node, but not every node, is close to high-degree nodes. Some nodes may require as many as $\mathcal{O}(\log N)$ steps to reach a high-degree node. See [CL01] for details.

Let us now consider the $l+1$ layer. There is a new threshold function, that is, the new distribution of unvisited nodes is like a step function – almost $P(k)$ for $k < K_{l+1}$ and almost 0 for $k > K_{l+1}$. The reason is as follows. A node with degree k has a probability of $p = k/(N\langle k\rangle)$ of being reached by following a link.[4] If there are χ_l outgoing links, then if $p\chi_l > 1$, we can assume that (in the limit $N \to \infty$) the node will be reached in the next level with probability 1. Therefore, all unvisited nodes with degree $k > N\langle k\rangle/\chi_l$ will definitely be reached in the next layer. On the other hand, almost all the unvisited nodes with degree $k < N\langle k\rangle/\chi_l$ will remain unvisited in the next layer. Therefore, their distribution will remain almost unchanged. It follows from those considerations that the highest degree of the unexplored nodes in the $l+1$ layer is determined by:

$$K_{l+1} \approx N\langle k\rangle/\chi_l\,. \tag{6.7}$$

In the lth layer, the number of loops, i.e., the number of links connecting two nodes of the lth layer and the number of nodes in the $l+1$ layer connected to more than one node in the lth layer is proportional to $\chi_l^2/(\langle k\rangle N)$. As long as χ_l is not of order N, this fraction is smaller in order than χ_l. Thus, we can safely assume that loops can be neglected until the last shells have been reached. Similar arguments have been used in [CEbH00].

In the $l+1$ layer, all nodes with degree $k > N\langle k\rangle/\chi_l$ will be exposed. Since the probability of reaching a node via a link is proportional to $kP(k)$, the average degree of nodes reached by following a link is $\kappa \equiv \langle k^2\rangle/\langle k\rangle$ (see Chapter 9). For scale-free graphs, κ can be approximated by [CEbH00],

$$\kappa = \left(\frac{\gamma-2}{\gamma-3}\right)\left(\frac{K^{3-\gamma}-m^{3-\gamma}}{K^{2-\gamma}-m^{2-\gamma}}\right)\,. \tag{6.8}$$

This will be the average degree for nodes reached in this layer, whose degree is $k < N\langle k\rangle/\chi_l$. Therefore, κ should be calculated using the new cutoff (6.7), from (6.8), which follows $\kappa \sim K_{l+1}^{3-\gamma}$.

Using the above considerations, we can calculate the number of outgoing links from the $l+1$ layer. To this end, we consider the total degree of all nodes reached in the $l+1$ level. This includes all nodes with degree k, $K_{l+1} < k < K_l$, as well as other nodes with an average degree proportional to $\kappa - 1$ links (the -1 is due to one link going inwards). Thus, the value of κ is calculated using the cutoff K_{l+1}. Loops within a layer and multiple links connecting the same node in the $l+1$ layer can be neglected, since as long as the number of nodes in the layer are of an order less than N, they are negligible in the limit $N \to \infty$. The two contributions can be written as

[4] We assume that $\langle k\rangle$ for the unvisited nodes is fixed since it is controlled by the low-degree nodes whose distribution is unchanged.

the sum of two terms:

$$\chi_{l+1} = N \int_{K_{l+1}}^{K_l} (k-1)P(k)\mathrm{d}k + \chi_l\,[\kappa(K_{l+1}) - 1]. \tag{6.9}$$

Using $P(k) \propto k^{-\gamma}$ and $\kappa \propto K^{3-\gamma}$ [CEbH00], it follows that $\chi_{l+1} \propto NK_{l+1}^{2-\gamma}$ (where both terms in Eq. (6.9) contribute the same order). This can be written as a second recurrence equation:

$$\chi_{l+1} = ANK_l^{2-\gamma}\,, \tag{6.10}$$

where $A = \langle k\rangle\, m^{\gamma-2}/(3-\gamma) = (\gamma-1)m/(\gamma-2)(3-\gamma)$.

Solving Eqs. (6.7) and (6.10) yields the result

$$\chi_l \sim A^{\frac{(\gamma-2)^l-1}{\gamma-3}} N^{1-\frac{(\gamma-2)^{l+1}}{\gamma-1}}, \tag{6.11}$$

where χ_l is the number of outgoing links from the lth layer. Equation (6.7) then leads to:

$$K_l \sim A^{\frac{(\gamma-2)^{l-1}-1}{3-\gamma}} N^{\frac{(\gamma-2)^l}{\gamma-1}}. \tag{6.12}$$

Using the same considerations that follow Eq. (6.6), one can deduce that here also

$$d \approx \ln\ln N/\ln(\gamma-2). \tag{6.13}$$

This result, Eq. (6.13), is consistent with the observations that the distance in the Internet network is extremely small and that the distance in metabolic scale-free networks is almost independent of N [JTA+00]. These results can be explained by the fact that $\ln\ln N$ is almost a constant over many orders of magnitude. The above arguments show, however, that for a fixed distribution and very large values of N, no scale-free graph with $\gamma > 2$ can have a constant diameter. However, for $\gamma = 2$, since the highest degree node has order N links, we expect that for this case $d \approx$ constant.

For $\gamma > 3$ and $N \gg 1$, κ is independent of N, and since the second term of Eq. (6.9) is dominant, Eq. (6.9) reduces to $\chi_{l+1} = (\kappa - 1)\chi_l$, where κ is a constant depending only on γ. This leads to the known result $\chi_l \approx C(N,\gamma)(\kappa-1)^l$ and the radius of the network is $l \propto \ln N$ [NSW01].

For $\gamma = 3$, Eq. (6.9) reduces to $\chi_{l+1} = \chi_l \ln \chi_l$. Taking the logarithm of this equation, one obtains $\ln\chi_{l+1} - \ln\chi_l = \ln\ln\chi_l$. Defining $g(l) = \ln\chi_l$ and approximating the difference equation yields a differential equation, $g' = \ln g$. This equation cannot be solved exactly. However, substituting $u = \ln g$, the equation reduces to

$$l = \int_{\ln\ln\sqrt{N}}^{\ln\ln N} \mathrm{e}^{u-\ln u}\mathrm{d}u. \tag{6.14}$$

The lower bound is obtained from the highest degree node for $\gamma = 3$, having degree $K = m\sqrt{N}$. Thus, $\chi_0 = m\sqrt{N}$. The upper bound is the result of searching l for which

$\chi_l \sim N$ with lower order corrections. The integral in Eq. (6.14) can be approximated by the steepest descent method, leading to

$$l \approx \ln N/(\ln \ln N), \tag{6.15}$$

for $\ln \ln N \gg 1$.

The above result, Eq. (6.15), was obtained rigorously for the maximum distance in the Barabási–Albert (BA) model [BA99], having $\gamma = 3$ (for $m \geq 2$) [BR02]. Although the result in [BR02] is for the largest distance between two nodes, their derivation makes it clear that the average distance will also behave similarly. For $m = 1$ in the Barabási–Albert model, the graph becomes a tree and the behavior of $d \sim \ln N$ is obtained [BR02, SAK02]. It should also be noted that for $m = 1$ the giant component in the random model contains only a fraction of the nodes (while for $m \geq 2$ it contains all the nodes, at least in the leading order). The BA model, on the other hand, is fully connected for every m. This might explain why exact trees and BA trees are different from generalized random graphs.

6.4 Layer structure and Internet tomography – how far do your emails travel?

In this section we describe the network structure by presenting the statistical properties of layers surrounding the maximal connected node. First, we describe the process of generating the network, and define our terminology. Then, we present the degree distribution at each layer surrounding the maximally connected node. The results presented here are based on [KCM$^+$06].

6.4.1 Description

We base our construction on the Bollobás model [Bol85]. The construction process tries to expose the network gradually, following the method introduced in the previous sections, enabling us to define layers in the graph.

We set the number of nodes in the network, N, and associate degrees with the nodes according to the scale-free distribution function $P(k) = ck^{-\gamma}$, where $c \approx (\gamma - 1)m^{\gamma-1}$ is the normalizing constant and k is in the range $[m, K]$, for some chosen minimal degree m and the natural cutoff $K = mN^{1/(\gamma-1)}$ of the distribution (see Section 4.3).

At this stage, each node in the network has a given number of outgoing links, which we term *open links* (or "stubs"), according to its chosen degree. Let us define V as the set of N chosen nodes, C as the set of unconnected outgoing links from the nodes in V, and E as the set of edges in the graph. Using these definitions, the set of links in E is empty at this point, whereas the set of outgoing open links in C contains all unconnected outgoing links in the graph. In the Bollobás construction [Bol80], the links in C are randomly matched, such that at the end of the process, C is empty, and E contains all the matched links $\langle u, v \rangle$, $u, v \in V$.

Instead, here we proceed as follows: we start from the maximal degree node, which has a degree K, and connect it randomly to K available open links, thus removing these open links from C (see Figure 6.3(a)). We have now exposed the first *layer* (or *shell*) of nodes, indexed as $l = 1$. We now continue to fill out the second layer $l = 2$ similarly. We connect all open links emerging from nodes in layer 1 to randomly chosen open links. These open links may be chosen from nodes of layer 1 (thus creating a loop) or from other links in C. We continue until all open links emerging from layer 1 have been connected, thus filling layer $l = 2$ (see Figure 6.3(b)). Generally, to form layer $l + 1$ from an arbitrary layer l, we randomly connect all open links emerging from l to either other open links emerging from l or chosen from the other links in C (see Figure 6.3(c)). Note that when we have formed layer $l + 1$, layer l has no more open links. The process continues until the set of open links, C, is empty.

6.4.2 Theory

We now proceed to evaluate the probability for nodes with degree k to reside outside the first l layers, denoted by $P_l(k)$. The number of open links outside layer l is given by:

$$T_l = N \sum_k k P_l(k) \,. \tag{6.16}$$

Thus, we can define the probability that a detached node with degree k will be connected to an open link emerging from layer l by $k/(\chi_l + T_l)$, where χ_l is the number of open links emerging from layer l (see Figure 6.3(b)).

Therefore, the conditional probability for a node with degree k to also be outside layer $l + 1$, given that it is outside layer l, is the probability that it does not connect to *any* of the χ_l open links emerging from layer l, that is:

$$P(k, l+1|l) = \left[1 - \frac{k}{\chi_l + T_l}\right]^{\chi_l} \approx \exp\left(-\frac{k}{1 + \frac{T_l}{\chi_l}}\right), \tag{6.17}$$

for large enough values of χ_l.

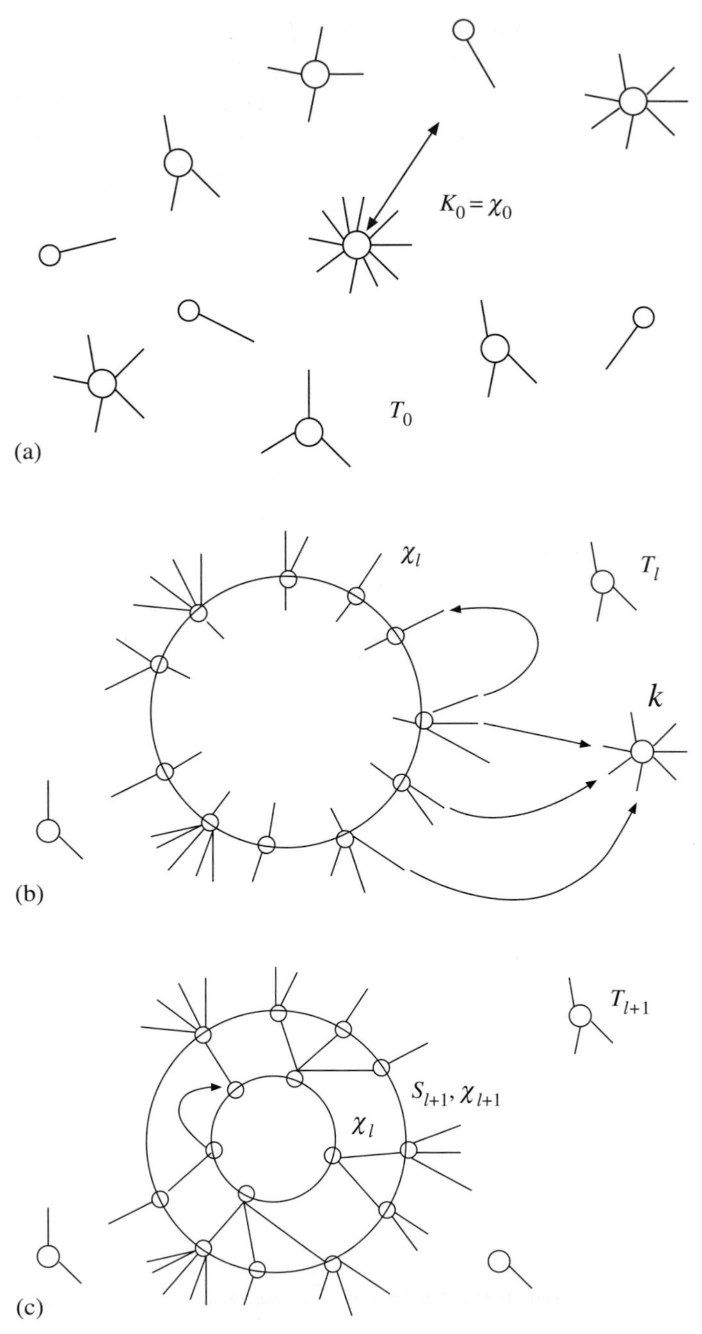

Figure 6.3 Illustration of the exposure process. The large circles denote exposed layers of the giant component, whereas the small circles denote individual nodes. The nodes outside the circles have not yet been reached. (a) We begin with the highest degree node and fill out layer 1. (b) In the exposure of layer $l + 1$ any open link emerging from layer l may connect to any open node (T_l links) or loop back into layer l (χ_l links). (c) The number of links emerging from layer $l + 1$ is the difference between T_l and T_{l+1} after reducing the incoming links S_{l+1} from layer l. After [KCM+06].

Thus, the probability that a node of degree k will be outside layer $l+1$ is:

$$P_{l+1}(k) = P_l(k)P(k, l+1|l) = P_l(k)\exp\left(-\frac{k}{1+\frac{T_l}{\chi_l}}\right). \tag{6.18}$$

Equation (6.18) yields the exponential cutoff:

$$P_l(k) = P(k)\exp\left(-\frac{k}{K_l}\right), \tag{6.19}$$

where

$$\frac{1}{K_{l+1}} = \frac{1}{K_l} + \frac{1}{1+\frac{T_l}{\chi_l}}, \tag{6.20}$$

gives the evolution of the cutoff with l.[5]

Now let us determine the behavior of χ_l and S_l, where S_{l+1} is the number of links incoming to the $l+1$ layer (and approximately[6] equals N_{l+1}, the number of nodes in the $l+1$ layer). The number of incoming links to layer $l+1$ equals the number of links emerging from layer l, minus the number of links looping back into layer l. The probability for a link to loop back into layer l is

$$P(\text{loop}|l) = \frac{\chi_l}{\chi_l + T_l} \tag{6.25}$$

[5] The exponential cutoff may also be derived using the following "mean-field" approximation. Each node is treated independently, where the *interaction* between nodes is inserted through the expected number of incoming links. At each node, the process is treated as equivalent to randomly distributing χ_l independent points on a line of length $\chi_l + T_l$ and counting the resulting number of points inside a *small* interval of length k. Thus, the number of incoming links k_{in} from layer l to a node with k open links is distributed according to a Poisson distribution with

$$\langle k_{\text{in}} \rangle = \frac{k}{\chi_l + T_l}\chi_l, \tag{6.21}$$

and

$$P_{l+1}(k_{\text{in}}|k) = \mathrm{e}^{-\langle k_{\text{in}} \rangle}\frac{\langle k_{\text{in}} \rangle^{k_{\text{in}}}}{k_{\text{in}}!}. \tag{6.22}$$

The probability for a node with k open links *not* to be connected to layer l, i.e., to also be outside layer $l+1$ is:

$$P(k, l+1|l) = P_{l+1}(k_{\text{in}} = 0|k) = \mathrm{e}^{-\langle k_{\text{in}} \rangle} = \exp\left(-\frac{k}{1+\frac{T_l}{\chi_l}}\right). \tag{6.23}$$

Thus, the total probability of finding a node of degree k outside layer $l+1$ is:

$$P_{l+1}(k) = P_l(k)P(k, l+1|l) = P_l(k)\exp\left(-\frac{k}{1+\frac{T_l}{\chi_l}}\right), \tag{6.24}$$

and one obtains the exponential cutoff.

[6] This holds true assuming that almost no node in layer $l+1$ is reached by two links from layer l. This is justified when $m = 1$, and also for the first layers if $m > 1$.

and therefore

$$S_{l+1} = \chi_l \left(1 - \frac{\chi_l}{\chi_l + T_l}\right) . \tag{6.26}$$

The number of links emerging from all the nodes in layer $l + 1$ is $T_l - T_{l+1}$. This is the sum of the number of incoming links from layer l into layer $l + 1$, which is equal to the sum of S_{l+1} and the number of outgoing links χ_{l+1},

$$\chi_{l+1} = T_l - T_{l+1} - S_{l+1} . \tag{6.27}$$

At this point we have the following relations: for $T_{l+1}(K_{l+1})$, Eq. (6.16) and Eq. (6.19), for $S_{l+1}(\chi_l, T_l)$, Eq. (6.26), for $K_{l+1}(K_l, \chi_l, T_l)$, Eq. (6.20), and for $\chi_{l+1}(T_l, T_{l+1}, S_{l+1})$, Eq. (6.27). These relations may be solved numerically.[7] Note that approximate analytical results for the limit $N \to \infty$ [CH03, CHb02, DMS03] are given in the previous sections.

6.4.3 Simulations

Figure 6.4 shows results from simulations (symbols) for the number of nodes in layer l, which, as can be seen, are in agreement with the analytical curves of S_l (lines). We can see that starting from a given layer $l = L \approx 5$, the number of nodes decays exponentially. The layer index L is expected to be related to the radius of the graph. It can be seen that S_l is a good approximation for the number of nodes at layer l. This is true in cases when only a small fraction of nodes in each layer l have more than one incoming link. An example of this case is when $m = 1$, so that most of the nodes in the network have only one link. Figure 6.5 shows the results for $P_l(k)$ with similar good agreement between simulations and theory. Note the exponential cutoff, which becomes stronger with l (i.e., K_l is a monotonically decreasing function of l).

It is important to note that the simulation results give the degree distribution for the giant percolation cluster,[8] whereas the analytical reconstruction gives the probability distribution for the whole graph, including the finite clusters. This may explain the difference in the probability distributions for lower degrees: many low-degree nodes are not connected to the giant percolation cluster and therefore the probability distribution derived from the simulation is smaller for low k values.

Similar results were found in real Internet maps and multicast trees [KCM$^+$06]. Deviations from theory, which were observed for the Internet, may be attributed to correlations in node degrees [New02a] and hierarchical structures [VPV02].

[7] We begin with $K_0 = K$ (the natural cutoff of the network), $\chi_0 = K_0 = K$, and $P_0(k) = ck^{-\gamma}$.

[8] That is, the nodes to which there is a path from the highest degree node. See Chapters 9 and 10 for further details.

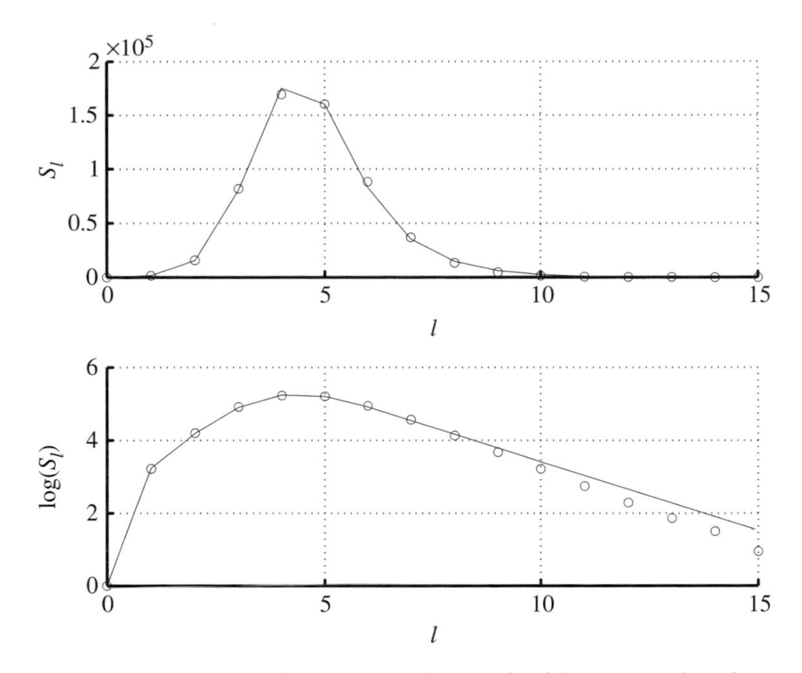

Figure 6.4 Top: approximate number of nodes, S_l, versus layer index l for a network with $N = 10^6$ nodes, $\gamma = 2.85$, and $m = 1$. Symbols represent simulation results, whereas solid lines are the numerical solution for the derived recursive relations, Eqs. (6.27) and (6.20). Bottom: from the semi-log plot, we see that there is an exponential decay of S_l for layers $l > L$ starting from a given layer $L \approx 5$, related to the radius of the graph. After [KCM+06].

6.4.4 Bounds and implications

The layer structure of the network has implications regarding several important topics. Since messages in a communication network travel between neighboring nodes, the internode distances are important in understanding network performance and message routing. Another important subject is searching for nodes in a network. In [ALPH01], an efficient method for searching via the network hubs is presented. This method is based on going up the degree sequence, from each node to a higher degree node, until the highest degree node is reached. Then, the search continues down the degree sequence to increasingly lower degree nodes. This method allows for a much more efficient search than a random one [ALPH01]. However, as we show below, no search strategy based on local information can search a finite fraction of the network in less than $\mathcal{O}(N)$ steps (with possible logarithmic corrections). (See Chapter 18 for further details.)

Some limits on the efficiency of such techniques can be obtained by using bounds on the structure of scale-free networks as in [CH03]. These bounds follow from

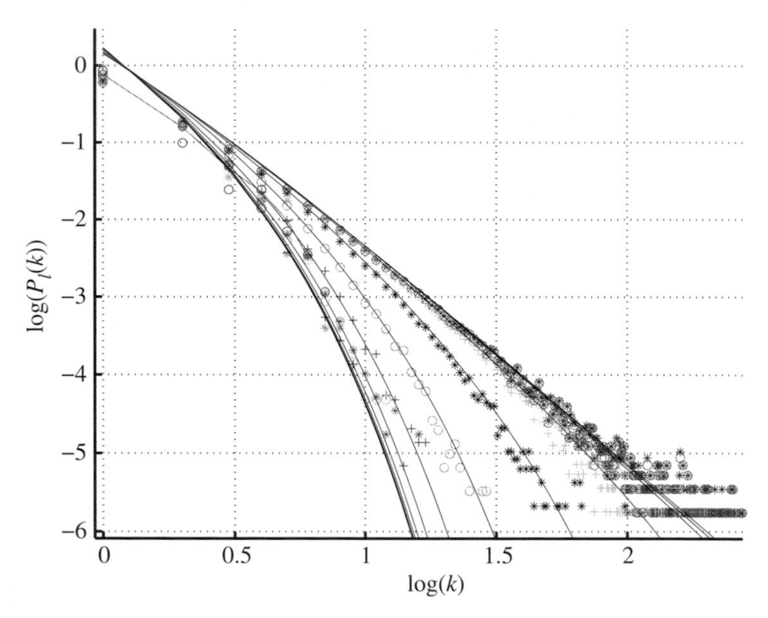

Figure 6.5 Log-log plot of $P_l(k)$ for different layers $l = 0, 1, 2, \ldots$ (from top to bottom), for a network with $N = 10^6$ nodes, $\gamma = 2.85$, and $m = 1$. Symbols represent simulation results, whereas solid lines represent a numerical solution for the derived recursive relations. After [KCM+06].

the scale-free nature of the network, and are independent of the exact model for network generation. Therefore, they apply to *every* network with a scale-free degree distribution.

If we consider the fraction, A, of the most highly connected nodes, their fraction of neighbors (relative to the network size), $n_1(A)$, cannot exceed $\int_f^K ck^{1-\gamma}\mathrm{d}k \approx f^{2-\gamma}m^{\gamma-1}(\gamma-1)/(\gamma-2)$, where f satisfies $N\int_f^K ck^{-\gamma}\mathrm{d}k = A$. Hence, $f = mA^{1/(\gamma-1)}$, and $n_1(A) \leq mA^{(\gamma-2)/(\gamma-1)}(\gamma-1)/(\gamma-2)$. Similarly, the number of next nearest neighbors of these nodes cannot exceed $n_2(A) \leq n_1(n_1(A)) \leq \left(m(\gamma-1)/(\gamma-2)\right)^{(2\gamma-3)/(\gamma-1)} A^{(\gamma-2)^2/(\gamma-1)^2}$. One can continue for the third nearest neighbors and so on.

Since the number of lth nearest neighbors of a group of AN nodes behaves as $NA^{((\gamma-2)/(\gamma-1))^l}$, it follows that for $A \sim N^{-\epsilon}$ for any ϵ, no finite number of layers can contain $\mathcal{O}(N)$ nodes in the limit $N \to \infty$. One can therefore conclude that the average distance between nodes in a scale-free network *cannot be a constant*. Also, *no searching of* $\mathcal{O}(N)$ *nodes using less than* $\mathcal{O}(N)$ *steps is possible*. The second conclusion is true, of course, only if no information other than lth nearest neighbors is allowed for some constant l, and does not apply to methods such as the one suggested in [WDN02], which can search in less than $\mathcal{O}(N)$ steps owing to knowledge of some distance metric. More details on searching in networks will be given in Chapter 18.

Note that there are different methods of defining a shell structure in networks. A straightforward generalization of the layer structure suggested above is the division into layers by the distance from some arbitrary node, rather than the highest degree one. A somewhat different division into shells is suggested in [CHK+07], where the network is divided into shells according to the k-core structure. See Section 10.6 for some details.

6.5 Discussion and conclusions

The above derivation is valid for uncorrelated networks. For assortative networks [New02a], the diameter is expected to be even smaller, as mentioned earlier. For disassortative networks, we would expect the odd layers to hold high-degree nodes and the even layers to hold low-degree nodes, so it is plausible that the scaling of the diameter is the same, with some possible constant factor ≤ 2. Note that this argument may not be valid for disassortative networks with $m = 1$, where many dead ends exist.

Note that there are other models for scale-free networks in which the distances are of order $d \sim \log N$, although the degree distribution is a power law with $\gamma < 3$. An example of such models is the tree construction presented in [Rb04] and the recursive construction in [JKK02].

In summary, we have seen that random scale-free graphs (with $2 < \gamma < 3$) have average diameter $d \sim \ln \ln N$, which is smaller than the $d \sim \ln N$ behavior expected for regular random graphs. For every $\gamma > 2$, scale-free graphs can be built to have a diameter of order $d \sim \ln \ln N$. If random scale-free graphs are considered only for $2 < \gamma < 3$, the behavior $d \sim \ln \ln N$ is obtained, whereas for $\gamma > 3$ the usual result $d \sim \ln N$ is obtained.

The results presented here, modified to take into account finite size effects, can be applied to moderately sized networks and multicast trees, used to broadcast information to many clients over the network. An additional study can be found in [KCM+06], and some of the methods will be presented in the following section.

The results presented here are based upon [CH03, CHb02]. Other methods for obtaining similar results can be found in [CL02, DMS03].

A different view of layers in the network is based upon k-core percolation, where layers are defined according to their shells in the process. In this way a Meduza model was proposed for the Internet network at the AS level. Details can be found in [CHK+07]. A relation between the degrees of two nodes, i and j, and the average distance between them, $d_{i,j} \sim A - B \ln(k_i k_j)$, can be found in [HSF+05]. For a more detailed study of the dependence of the node to node distance on the degree, see

[DMO06]. A discussion of the structure and properties of the boundary of a network, i.e., of nodes much farther than the average diameter may be found in [SBC$^+$08].

Exercises

6.1 What is the average distance in the model presented in Exercise 4.5? Compare this result to a random network with a similar degree distribution.

6.2 What is the average distance between nodes in a random regular graph with a constant degree k?

6.3 (a) What is the average distance between nodes in a random network with half the nodes having degree 3 and half the nodes having degree 5?

(b) What is the average distance between nodes in a maximally assortative network where half of the nodes have degree 3 and half have degree 5, i.e., where all nodes of degree 5 form a tree and all nodes of degree 3 are at the periphery?

(c) Where is the expected average distance higher, in an assortative or a disassortative network?

7 Self-similarity in complex networks

As has been seen throughout this book, two fundamental properties of real complex networks have attracted much attention recently: the small-world and the scale-free properties. Many naturally occurring networks are small world since one can reach a given node from another one, following the path with the smallest number of links between the nodes, in a very small number of steps. This corresponds to the so-called "six degrees of separation" in social networks [Mil67]. It is expressed mathematically by the slow (logarithmic or lower) increase in the average distance in the network, $\bar{l}$, with the total number of nodes N, $\bar{l} \sim \ln N$ (or, for scale-free networks with $2 \leq \gamma \leq 3$, the even smaller radius $\bar{l} \sim \ln \ln N$, as seen in Chapter 6), where $\bar{l}$ is the mean shortest distance between two nodes and defines the distance metric in complex networks. Thus, for exponential networks, we obtain:

$$N \sim e^{\bar{l}/l_0} \tag{7.1}$$

where l_0 is a characteristic length.

A second fundamental property in the study of complex networks arises from the discovery that for many real-world networks, the degree distribution, $P(k)$, can be represented by a power law (scale-free) with a degree exponent γ usually in the range $2 < \gamma < 3$ (see Chapter 3),

$$P(k) \sim k^{-\gamma} \ . \tag{7.2}$$

In aiming to provide a deeper understanding of the underlying mechanism that leads to these common features, one needs to probe the patterns within the network structure in more detail. The question of connectivity between groups of interconnected nodes on different length-scales has received less attention. Yet a plethora of examples in nature confirms the importance of collective behavior, from interactions between communities within social networks, links between clusters of websites of similar subjects, all the way to the highly modular manner in which molecules interact to keep a cell alive. Here we review the results of Song *et al.* [SHM05], showing that some real complex networks, such as the WWW, protein–protein interaction networks (PIN) and cellular networks are indeed constructed of self-repeating patterns on all length-scales, and are therefore invariant or self-similar under a length-scale transformation.

This result is surprising since the exponential increase in Eq. (7.1) has led to the general understanding that complex networks are not self-similar, since self-similarity requires a power-law relation between N and l [BH94, bH00, Fed88].

In order to demonstrate this concept, let us consider a self-similar network embedded in Euclidean space, of which a classical example would be a fractal percolation cluster at criticality [BH96, SA94]. The self-similar property of such clusters can be observed by measuring the fractal dimension using a "box counting" method and a "cluster growing" method [BH96].

In the box counting method, we cover the percolation cluster with N_B boxes of size (minimum distance) l_B. The fractal dimension or box dimension d_B is then given by [BH96, Fed88]:

$$N_B \sim l^{-d_B} . \tag{7.3}$$

In the cluster growing method, the network is not covered with boxes; instead, one seed node is chosen at random and the number of nodes ("mass") in a cluster centered at the seed and separated by a minimum distance l is calculated. The procedure is then repeated by choosing many seed nodes at random and the average "mass," $\langle M_c \rangle$, of the resulting clusters is calculated as a function of l to obtain the following scaling.

$$\langle M_c \rangle \sim l^{d_l} , \tag{7.4}$$

defining the fractal cluster dimension d_l [BH96].[1] Comparing Eqs. (7.4) and (7.1) suggests that $d_l = \infty$ for complex small-world networks. For homogeneous networks characterized by a narrow degree distribution (such as a percolation cluster at criticality)[2] the box-covering method of Eq. (7.3) and the cluster-growing method of Eq. (7.4) are equivalent since every node typically has the same number of links or neighbors. Equation (7.4) can then be derived from (7.3), ($\langle M_c \rangle = N_L/N_B = l^{d_B}$) and $d_B = d_l$. Both relations have been regularly used to determine the fractal dimension. In complex heterogeneous networks with a broad degree distribution such as in Eq. (7.2), Eqs. (7.3) and (7.4) may not be equivalent, as will be shown below. Applying the proper box counting method, Eq. (7.3), for complex networks reveals a set of self-similar properties such as a finite fractal dimension and a new set of critical exponents for the scale-invariant topology, whereas the cluster growing method reveals the small-world property of the networks.

Figure 7.1(a) illustrates the box covering method using a schematic network composed of 8 nodes. For each value of the box size l, we search for the minimal number

[1] Note that usually the fractal dimension, d_f, is defined based on the dependence of the mass on the radius in Euclidean space. Since networks are usually not assumed to be embedded in Euclidean space, the chemical dimension, d_l, which is the fractal dimension with respect to the metric of hop distances on the network, is used.

[2] See Chapter 10.

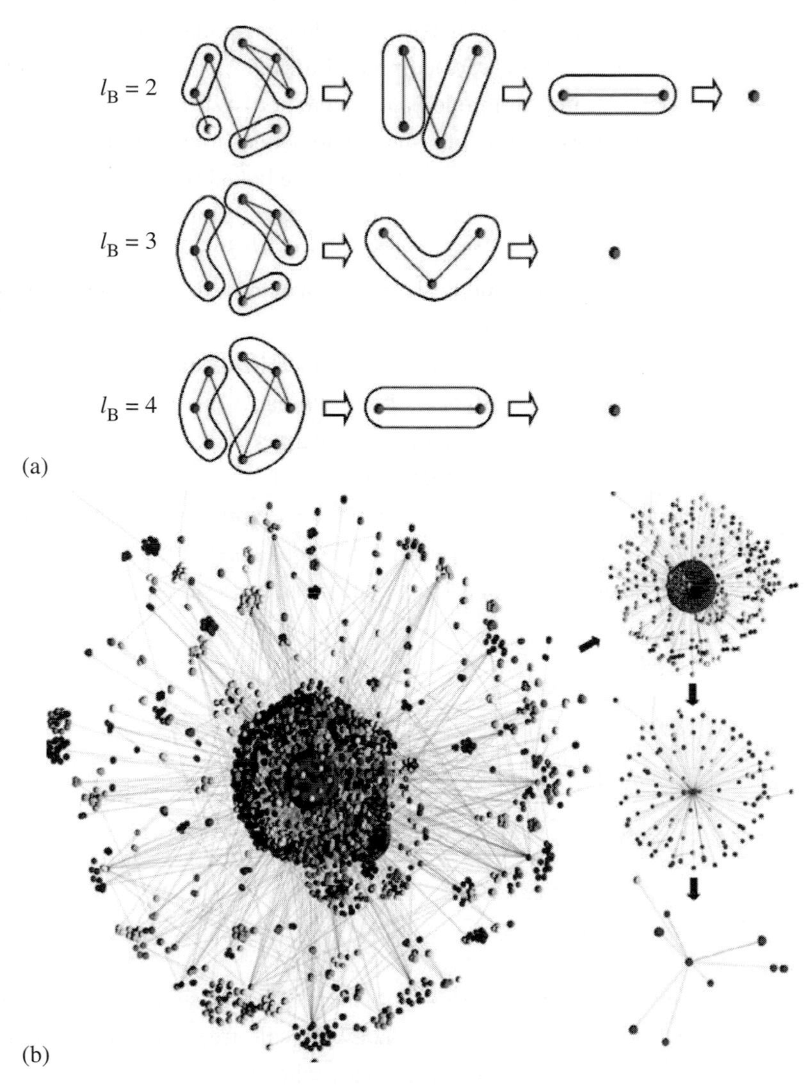

Figure 7.1 The renormalization procedure for complex networks. (a) Demonstration of the method for different l_B values and different stages in a network demo. The first column depicts the original network. We tile the system with boxes of size l_B. All nodes in a box are connected by a minimum distance smaller than the given l_B. For instance, in the case of $l_B = 2$, we identify four boxes that contain the nodes, each containing 3, 2, 1, and 2 nodes, respectively. Then we replace each box by a single node; two renormalized nodes are connected if there is at least one link between the unrenormalized boxes. Thus, we obtain the network shown in the second column. The resulting number of boxes needed to tile the network, $N_B(l_B)$, is plotted in Figure 7.2 versus l_B to obtain d_B as in Eq. (7.3). The renormalization procedure is applied again and repeated until the network is reduced to a single node (third and fourth columns for different l_B values). (b) Three stages in the renormalization scheme applied to the entire WWW. We fix the box size to $l_B = 3$ and apply the renormalization for four stages. This corresponds, for instance, to the sequence for the network demo depicted in the second row in part (a) of this figure. The network is invariant under this renormalization, as explained in the legend of Figure 7.2(d). After [SHM05].

of boxes needed to tile the entire network such that each box contains nodes separated by a distance $\leq l$. This procedure is applied to several different real networks: (i) a part of the WWW composed of 325 729 webpages that are connected if there is a URL link from one page to another [AJB99] (see Figure 7.1(b)), (ii) the biological network of protein–protein interactions found in *E. coli* (429 proteins) and *H. sapiens* (human) (946 proteins), which are linked if there is a physical binding between them (database available via the Database of Interacting Proteins [XSD$^+$02]). It has been previously determined that the WWW and the PINs of *E. coli* and *H. sapiens* are small world and scale free, characterized by Eq. (7.2) with $\gamma = 2.6$, 2.2, and 2.1, respectively [AB02].

Figures 7.2(a) and 7.2(b) show the results of $N_{\mathrm{B}}(l)$ according to Eq. (7.3). They reveal the existence of self-similarity in the WWW and *E. coli* and *H. sapiens* protein–protein interaction networks with self-similar exponents $d_{\mathrm{B}} = 4.1$, $d_{\mathrm{B}} = 2.3$ and $d_{\mathrm{B}} = 2.3$, respectively.

We will now elaborate on the seeming contradiction between the two definitions of self-similar exponents in complex networks. After performing a renormalization at a given l, we calculate the mean mass of the boxes covering the network, $\langle M_{\mathrm{B}}(l)\rangle$, to obtain

$$\langle M_{\mathrm{B}}(l)\rangle \equiv N/N_{\mathrm{B}}(l) \sim l^{d_{\mathrm{B}}}, \tag{7.5}$$

which is corroborated by direct measurements for all the networks and is shown in Figure 7.3(a) for the WWW. On the other hand, the average performed in the cluster growing method (for this calculation we average over single boxes without tiling the system) gives rise to an exponential growth of the mass

$$\langle M_{\mathrm{c}}(l)\rangle \sim \mathrm{e}^{l/l_1}, \tag{7.6}$$

with $l_1 \approx 0.78$ in accordance with the small-world effect, Eq. (7.1), as seen in Figure 7.3(a). The topology of scale-free networks is dominated by several highly connected hubs – the nodes with the largest degree – implying that most of the nodes are connected to the hubs via one or a very few steps. Therefore, the average performed in the cluster growing method is biased with respect to determining the fractal dimension; the hubs are overrepresented in Eq. (7.6) since almost every node is a neighbor of a hub. Therefore, this method is not appropriate for determining the fractal dimension, but determines well the typical distance between nodes, the small-world property. On the other hand, the box covering method involves global tiling of the system, providing an average over all the nodes, i.e., each part of the network is covered with the same probability. Thus, Eqs. (7.3) and (7.4) are not equivalent for inhomogeneous networks with topologies dominated by hubs with a large degree.

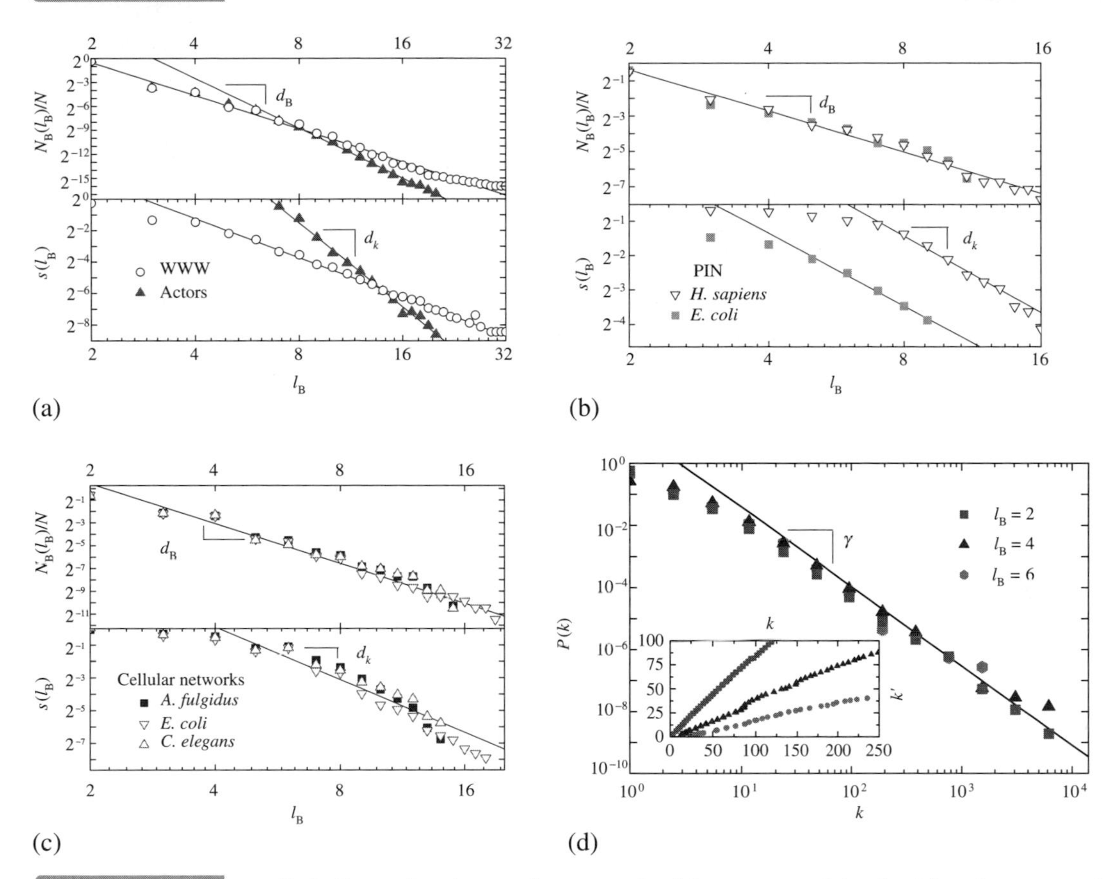

Figure 7.2 Self-similar scaling in complex networks. (a) Upper panel: log-log plot of N_B versus l_B revealing the self-similarity of the WWW and actor network according to Eq. (7.3). Lower panel: the scaling of $s(l_B)$ versus l_B according to Eq. (7.9). The error bars are of the order of the symbol size. (b) Same as (a) but for two protein interaction networks: *H. sapiens* and *E. coli*. Results are analogous to (a) but with different scaling exponents. (c) Same as (a) for the cellular networks of *A. fulgidus*, *E. coli* and *C. elegans*. (d) Invariance of the degree distribution of the WWW under the renormalization for different box sizes, l_B. We show the data collapse of the degree distributions, demonstrating the self-similarity at different scales. The inset shows the scaling of $k' = s(l_B)k$ for different l_B values, where we obtain the scaling factor $s(l_B)$. Moreover, we also apply the renormalization for a fixed box size, for instance, $l_B = 3$, as shown in Figure 7.1(b) for the WWW, until the network is reduced to a few nodes. We found that $P(k)$ is invariant under these multiple renormalizations as well, for several iterations. After [SHM05].

The box covering method serves as a powerful tool for further investigating the network properties, since it enables a renormalization procedure, revealing that the self-similar properties and the scale-free degree distribution persist after coarse graining of the network.

After the first step of assigning the nodes to the boxes, one creates a new renormalized network by replacing each box by a single super node. Two boxes are connected

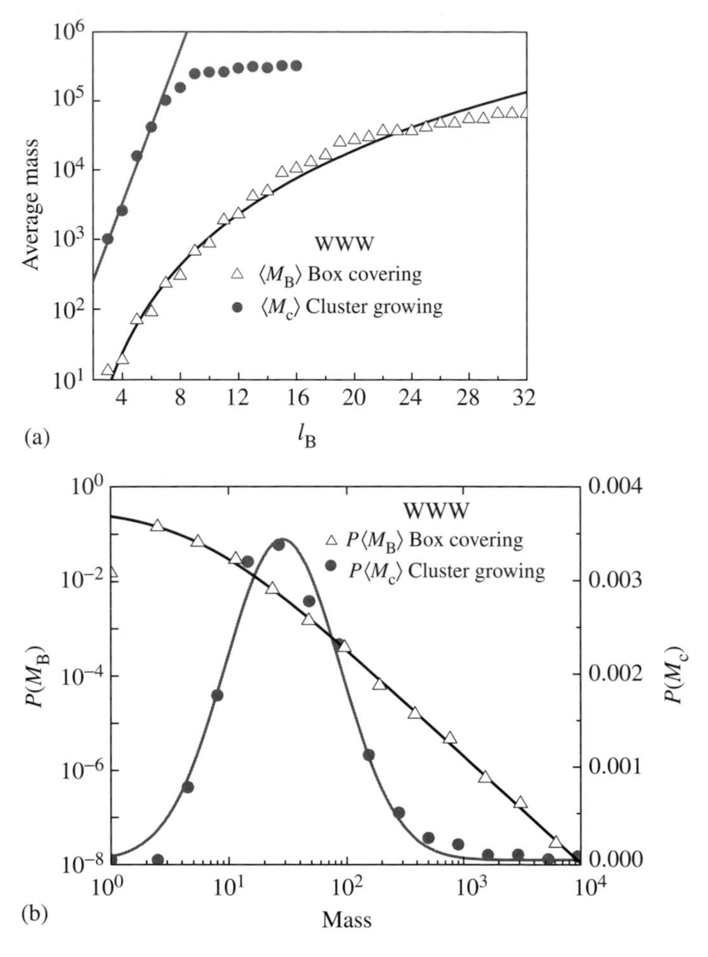

Figure 7.3 Different averaging techniques lead to qualitatively different results. (a) Mean value of the box mass in the box counting method, $\langle M_B \rangle$, and the cluster mass in the cluster growing method, $\langle M_c \rangle$, for the WWW. The solid lines represent the power-law fit for $\langle M_B \rangle$ and the exponential fit for $\langle M_c \rangle$ according to Eqs. (7.5) and (7.6), respectively. (b) Probability distribution of M_B and M_c for $l_B = 4$ for the WWW. The curves are fitted by a power-law and a log-normal distribution, respectively. After [SHM05].

if there was at least one link between their constituent nodes.[3] The second column of the panels in Figure 7.1(a) shows this step in the renormalization procedure for the schematic network, whereas Figure 7.1(b) shows the results for the same procedure applied to the entire WWW for $l_B = 3$. The renormalized network gives rise to a new probability distribution of links, $P(k')$, which is invariant under the renormalization:

$$P(k) \rightarrow P(k') \sim (k')^{-\gamma}. \tag{7.7}$$

[3] Note that one can assign weights to the links between supernodes according to the number of links between the boxes.

Figure 7.2(d) supports the validity of this scale transformation by showing a data collapse of all distributions with the same γ according to Eq. (7.7) for the WWW. Further insight is gained from relating the scale-invariant properties (7.3) to the scale-free degree distribution (7.7). Plotting (see inset in Figure 7.2(d) for the WWW) the number of links k' of each node in the renormalized network versus the maximum number of links k in each box of the unrenormalized network reveals a scaling law

$$k \rightarrow k' = s(l)k. \tag{7.8}$$

This equation defines the scaling transformation in the connectivity distribution. Empirically, we see that the scaling factor s ($s < 1$) scales with l with a new exponent d_k as

$$s(l) \sim l^{-d_k}, \tag{7.9}$$

shown in Figure 7.2(a) for the WWW network (with $d_k = 2.5$).

Equations (7.8) and (7.9) shed light on how families of hierarchical sizes are linked together. The larger the families, the fewer are the links between them. Surprisingly, the same power-law relation exists for large and small families represented by Eq. (7.2).

From Eq. (7.7) we obtain $n(k)dk = n'(k')dk'$, where $n(k) = NP(k)$ is the number of supernodes with k links, and $n'(k') = N'P(k')$ is the number of nodes with k' links after the renormalization (N' is the total number of supernodes in the renormalized network). Using Eq. (7.8), we obtain $n(k) = s^{1-\gamma}n'(k)$. Then, upon renormalizing a network with N total nodes, we obtain a smaller number of nodes N' according to $N' = s^{\gamma-1}N$. Since the total number of nodes in the renormalized network is the number of boxes needed to cover the unrenormalized network at any given l, we have $N' = N_{\mathrm{B}}(l)$. Hence, from Eqs. (7.3) and (7.9), we obtain the relation between the three indices:

$$\gamma = 1 + d_{\mathrm{B}}/d_k. \tag{7.10}$$

The significance of this result is that the scale-free property characteristics can be related to a more fundamental length-scale invariant property, characterized by the two new exponents d_{B} and d_k.

It has been shown [SGHM07] that the fractal properties of networks may be explained by repulsion between hubs. Methods for generating fractal networks with scale-free degree distribution have been studied in [GSKK06, RHb07].

Further results on self-similarity in networks may be found in [SHM06]. A discussion on the fractal skeleton of networks may be found in [GSKK06]. It has also been demonstrated that the boundaries of networks, i.e., the nodes most distant from a single node, present fractal properties, see [SBC+08].

8 Distances in geographically embedded networks

We now address the geometrical properties of the networks, arising from their embedding in Euclidean space (see Section 4.6). For this purpose, it is useful to consider the spatial arrangement of the networks as measured both in a Euclidean metric and in *chemical space*. The chemical distance l between any two nodes is the minimal number of links between them (*shortest path*). Thus, if the distance between the two nodes is r, then $l \sim r^{d_{\min}}$ defines the shortest path exponent $d_{\min}$. We will see that for scale-free networks embedded in $d > 1$ lattices, $d_{\min} < 1$, contrary to all known fractals and disordered media where $d_{\min} \geq 1$. Nodes at chemical distance l from a given node constitute its lth chemical shell. Thus, the number of (connected) nodes within radius r scales as $m(r) \sim r^{d_f}$, defining the fractal dimension d_f. Likewise, the number of (connected) nodes within chemical (hop) radius l scales as $m(l) \sim l^{d_l}$, which defines the fractal dimension d_l in chemical (hop) space.[1] The two fractal dimensions are related: $d_{\min} = d_f/d_l$ [BH96, bH00, SA94]. Note that $d_{\min}$ has a meaning only when the network is embedded in Euclidean space. In networks that are not embedded, the relation between $M(l)$ and l exists, but it can only be a power law when the networks are fractals (Chapter 7) or at the percolation threshold (Chapter 10). It can be seen that the small-world property of complex networks disappears when the networks are embedded in Euclidean space. Although the distances are still smaller than in regular lattices, they are much larger than in non-embedded networks. This is due to the restriction of the maximal distance between neighboring nodes.

To study d_f, we compute the perimeter $S(r)$, the number of nodes that connect the interior cluster of a region of radius r to nodes outside. The fractal dimension then follows from the scaling relation $S(r) \sim r^{d_f-1}$. We focus on the regime $\xi > r_{\max}$. Consider a shell dr', of radius r'. A node of connectivity k' within the shell is connected to the outside (to a distance larger than $r - r'$) with probability $P(k' > (\frac{r-r'}{A})^d)$.

[1] Note that the fractal dimension d_B discussed in Chapter 7 is identical to d_l in chemical (hop) space since no embedding in Euclidean space is enforced.

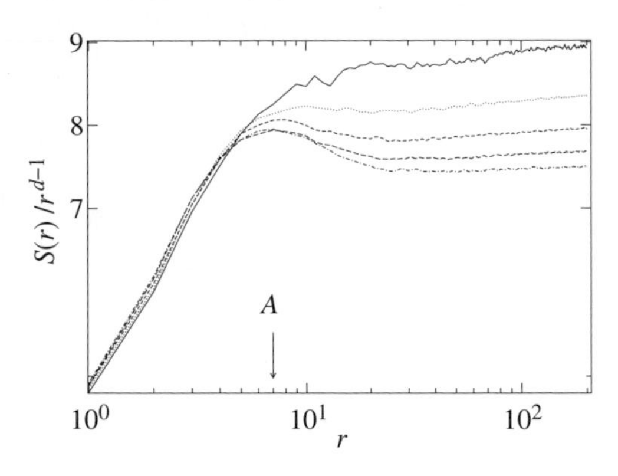

Figure 8.1 Plot of a scaled perimeter as a function of the Euclidean distance from the central node, for several values of γ: $\gamma = 3.0$ (top), 3.25, 3.5, 3.75 and 4.0 (bottom). The simulations were performed with $A = 7$. Note that the position where the curves split, $r \simeq A$, is consistent with our analytical results (Eq. (8.1)). Also, the asymptotic values shown for large r values are consistent with $c(\gamma)A$. After [RCbH02].

Thus,

$$S(r) = \int_0^r \mathrm{d}r' r'^{d-1} P\left(k' > \left(\frac{r-r'}{A}\right)^d\right) \sim \begin{cases} r^d & r < A \\ c(\gamma)Ar^{d-1} & r > A, \end{cases} \tag{8.1}$$

where $c(\gamma) \sim 1 + 1/[d(\gamma - 1) + 1]$. In other words, the network is compact, $d_{\mathrm{f}} = d$ at large distances $r > A$, and the network is super-compact, $d_{\mathrm{f}} = d + 1$ at $r < A$. Results for $S(r)$ are presented in Figure 8.1 and are in good agreement with Eq. (8.1). The slight negative slope observed for $r \gtrsim A$ is due to analytical corrections, of order r^{-1}, to the scaling $S(r) \sim r^{d-1}$, and can be derived from a more careful analysis of Eq. (8.1).

In order to compute $d_{\min}$ (or d_l), we regard the chemical shells as being roughly smooth, at least in the regime $\xi > r_{\max}$. Let the width of shell l be $\Delta r(l)$, then

$$l = \int \mathrm{d}l = \int \frac{\mathrm{d}r}{\Delta r(l)} \sim r^{d_{\min}}, \tag{8.2}$$

since $\Delta l = 1$. The number of nodes in shell l, $N(l)$, is, on the one hand, $N(l) \sim r(l)^{d-1}\Delta r(l)$. On the other hand, since the maximal connectivity in shell l is $K(l) \sim N(l)^{1/(\gamma-1)}$, the width of shell $(l+1)$ is $\Delta r(l+1)$, which is determined by the length of the largest link to the next shell, i.e., $r[K(l)]$, and thus, $\Delta r(l+1) \sim r[K(l)] \sim AK(l)^{1/d}$. Assuming (for large l values) that $\Delta r(l+1) \sim \Delta r(l)$, we obtain

$$\Delta r(l) \sim r^{\frac{d-1}{d(\gamma-1)-1}}. \tag{8.3}$$

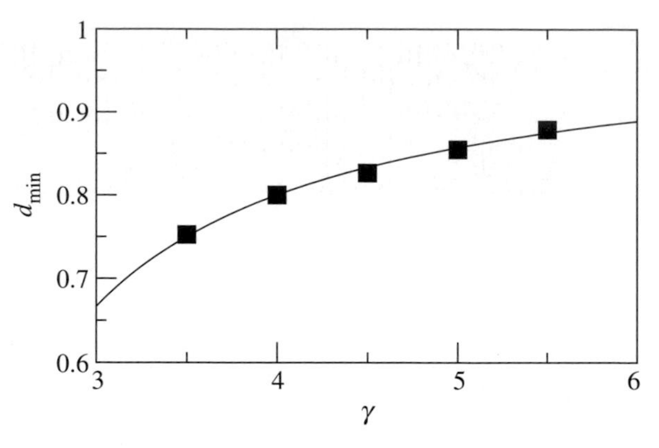

Figure 8.2 The minimal path length exponent d_{min} as a function of γ. Note the good agreement between theoretical estimations, Eq. (8.4), (continuous line) and simulation results (full squares). After [RCbH02].

Using this expression in (8.2) yields

$$d_{min} = \frac{\gamma - 2}{\gamma - 1 - 1/d}. \tag{8.4}$$

Thus, above $d = 1$, the fractal dimensions d_{min} and $d_l = d_f/d_{min}$ are anomalous for all values of γ.

In Figure 8.2 we plot d_{min}, as measured from simulations, and compare the result with the analytical result Eq. (8.4). The scaling $N(l) \sim l^{d_l-1}\Phi(l^{d_l}/R^d)$, is valid only for $\xi > r_{max}$. For $R \to \infty$, we expect that the network will be scale free up to length-scale ξ and the analogous scaling will be $N(l) \sim l^{d_l-1}\Psi(l^{d_l}/\xi^d)$, where $\Psi(x \gg 1) \sim x^{(d-d_l)/d_l}$.

9 The structure of networks: the generating function method

9.1 Introduction

A generalized random network is generally made up of several components. In the random generalized graph model, the structure of the network is determined only by the degree distribution, since no other structure or constraint is imposed on the network. In general, for many degree distributions the network is composed of many separate **components**, i.e., groups of nodes connected internally, but disconnected from other components. In each such component there exists a path between any two nodes, but there is no path between nodes in different components. When there is a component with size (the number of nodes) proportional to that of the entire network, it is called the **giant component** (sometimes also referred to as the infinite component or giant cluster or percolating cluster).

This chapter introduces a method for determining the component structure of a network, which is one of its most important structural properties. We will present a method for determining the existence of a giant component, and also a method for determining the sizes of all clusters using generating functions. The generating function method is a general and useful method for many probabilistic and combinatorial problems. An introduction to this method is given in Appendix A. The methods introduced in this chapter will be used in Chapter 10 to determine the robustness properties of networks and later in Chapters 14 and 15 to determine how epidemics propagate through networks and how they can be prevented by immunization.

9.2 General results

9.2.1 Condition for the existence of a giant component

For a graph having degree distribution $P(k)$ to have a giant component, a node that is reached by following a link from the giant component must have at least one other link on average to allow the component to exist. For this to occur, the average degree of a node must be at least 2 (one incoming and one outgoing link) given that the node i is connected to j:

$$\langle k_i | i \leftrightarrow j \rangle = \sum_{k_i} k_i P(k_i | i \leftrightarrow j) = 2. \tag{9.1}$$

Using Bayes' rule, we obtain

$$\begin{aligned} P(k_i | i \leftrightarrow j) &= P(k_i, i \leftrightarrow j) / P(i \leftrightarrow j) \\ &= P(i \leftrightarrow j | k_i) P(k_i) / P(i \leftrightarrow j), \end{aligned}$$

where $P(k_i, i \leftrightarrow j)$ is the *joint* probability that node i has degree k_i and that it is connected to node j. For randomly connected networks (neglecting loops) $P(i \leftrightarrow j) = \langle k \rangle / (N-1)$ and $P(i \leftrightarrow j | k_i) = k_i / (N-1)$, where N is the total number of nodes in the network. Using the above criteria, Eq. (9.1) reduces to [CEbH00]:

$$\kappa \equiv \frac{\langle k^2 \rangle}{\langle k \rangle} = 2, \tag{9.2}$$

at the critical point. A giant component exists for graphs with $\kappa > 2$, whereas graphs with $\kappa < 2$ contain only small components whose size is not proportional to that of the entire network. This criterion, Eq. (9.2), was also derived by Molloy and Reed [MR95] using somewhat different arguments.

The neglecting of loops can be justified below the threshold. The probability for a bond to form a loop in an s-node component is proportional to $(s/N)^2$ (i.e., proportional to the probability of choosing two nodes in that component). Thus, the fraction of loops P_{loop} in the system is:

$$P_{\text{loop}} \propto \sum_i \frac{s_i^2}{N^2} < \sum_i \frac{s_i S}{N^2} = \frac{S}{N}, \tag{9.3}$$

where the sum is over all components in the system, S is the size of the largest component, and s_i is the size of the ith component. Therefore, the fraction of loops in the system is less than or proportional to S/N. Below the critical threshold, there is no spanning cluster in the system and therefore the fraction of loops is negligible. Hence, until $\kappa = 2$, loops can be neglected. At the threshold, as will be shown later (Section 10.4.5), $S \sim N^{2/3}$, $P_{\text{loop}} \to 0$, and therefore the structure of the largest component is almost a tree. Above the threshold, loops can no longer be neglected, but since this only occurs when a giant component exists, the criterion in Eq. (9.2) is valid as a criterion for finding the critical point. A derivation of the exact conditions under which Eq. (9.2) is valid can be found in [MR95].

9.2.2 Generating functions

Molloy and Reed [MR98] were the first to develop a general method for studying the size of the giant component and the residual network for a graph with an arbitrary degree distribution. They suggested viewing and exploring the giant component, and used differential equations for the number of unexposed links and unvisited nodes to find the size of the giant component and the degree distribution of the residual graph (the finite components).

An alternative and very powerful approach was developed by Newman, Watts, and Strogatz [NSW01]. They used the generating function method to study the size of the giant component as well as other quantities (such as the diameter and component size distribution). They also applied this method to other types of graphs (directed and bipartite). Here we closely follow their derivation in order to find the size of the giant component. In Chapter 10 we will continue using this method in order to determine the size of the percolation components and the critical exponents of the percolation transition.

In [NSW01] a generating function is built for the degree distribution:

$$G_0(x) = \sum_{k=0}^{\infty} P(k)x^k. \tag{9.4}$$

The probability of reaching a node with degree k by following a specific link is $kP(k)/\langle k\rangle$ [CEbH00, CNSW00, MR95, NSW01], and the corresponding generating function for those probabilities is

$$G_1(x) = \frac{\sum kP(k)x^{k-1}}{\sum kP(k)} = \frac{\mathrm{d}}{\mathrm{d}x}G_0(x)/\langle k\rangle \ . \tag{9.5}$$

Assuming that $H_1(x)$ is the generating function for the probability of reaching a branch of a given size by following a link, the self-consistent equation for $H_1(x)$ is (see Figure 9.1):

$$H_1(x) = xG_1(H_1(x)) \ . \tag{9.6}$$

Since $G_0(x)$ is the generating function for the degree of a node, the generating function for the probability that a node belongs to an n-node component is:

$$H_0(x) = xG_0(H_1(x)) \ . \tag{9.7}$$

Below the transition, $H_0(1) = 1$, since this is the probability of belonging to a component of any size. However, above the transition this probability is no longer normalized since this does not include the giant component. Then, $P_\infty = 1 - H_0(1)$, since H_0

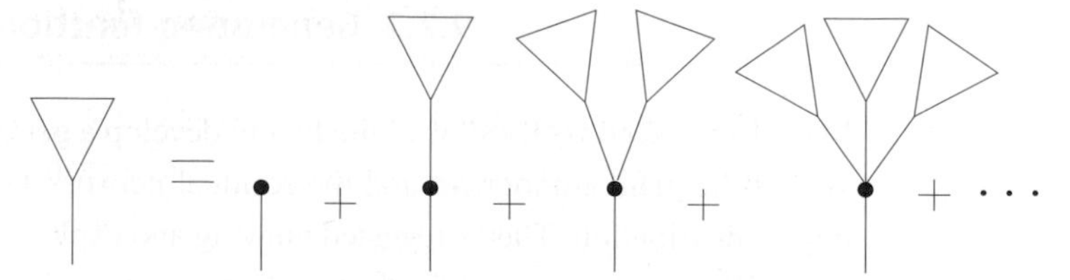

Figure 9.1 Illustration of the recursive relation for $H_1(x)$. A branch can consist of a single vertex, or a vertex connected to an arbitrary number of other branches. After [NSW01].

contains only the finite-size components. It follows that

$$P_\infty(p) = 1 - \sum_{k=0}^{\infty} \tilde{P}(k) u^k \,, \tag{9.8}$$

where $u \equiv H_1(1)$ is the smallest positive root (found numerically) of

$$\langle k \rangle \, u = \sum_{k=0}^{\infty} k P(k) u^{k-1} \,. \tag{9.9}$$

This equation can be solved numerically and the solution can be substituted into Eq. (9.8) to calculate the size of the giant component in a graph with a given degree distribution.

9.3 Scale-free networks

9.3.1 Description

One of the main concerns in this book is the structural properties of scale-free networks, with the degree distribution given by Eq. (2.2). There are no nodes with a degree below m and above K. For finite networks, the upper cutoff, K, arises naturally since the fraction of high-degree nodes decays with k and is given by $K \approx m N^{1/(\gamma-1)}$ (Eq. (2.4)).

A few results about the structure of scale-free networks have also been derived by Aiello *et al.* [ACL00]. The size of the giant component was calculated, and it was found that for $\gamma \leq 2$ the giant component is of the order of the size of the entire graph (i.e., $P_\infty = 1 - o(1)$, where $o(1)$ is a function of the network size, $f(N)$, such that $f(N) \to 0$ when $N \to \infty$). For $m = 1$ and $\gamma > \gamma_c = 3.478\ldots$, there is no giant component at all (since we use here a somewhat different distribution, Eqs. (3.2) and (3.3), we get $\gamma_c \approx 4$). For $\gamma < \gamma_c$, the second largest component is of order $\ln N$. If the lower cutoff is $m \geq 2$, a giant component exists for every γ.

9.3.2 Maximum degree

In Eq. (2.4) we suggest that the upper cutoff of a scale-free network scales as $N^{1/(\gamma-1)}$. However, for the spatially embedded graphs [RCbH02], we found that no graph with $\gamma < 3$ can be embedded in a lattice without sacrificing the natural cutoff. That is, the cutoff is limited to $\sqrt{N}$. This holds true for every d. Thus, we expect this to hold true even in the $d \to \infty$ case. Similar results are indeed obtained for the mean-field (i.e., non-embedded) graphs [BK03], whereas Warren *et al.* [WSS02b] found a natural cutoff even for graphs embedded in $d = 2$ lattices.

To explain this, it should be noted that the cutoff depends on the ensemble from which the graphs are chosen. If the ensemble is defined as all graphs with the exact given scale-free degree distribution and no self loops (loops connecting a node to itself) and no double edges (two edges or more connecting the same set of nodes), then the upper cutoff cannot be larger than $\sqrt{N}$, whereas if multigraphs are allowed (including self loops and double edges), the natural upper cutoff is achieved.

Consider a graph with a node having degree $K \gg \sqrt{N}$. Since for $\gamma > 2$ the total number of links is of order N, the number of self loops is proportional to $K^2/cN \gg 1$. Similar results apply for double edges between two such nodes. Thus, almost all graphs with the natural cutoff are multigraphs.

Suppose now that a node has degree $K \gg \sqrt{N}$. The number of degree 1 nodes is of order N. Therefore, the number of connections from this node to nodes of degree 1 is proportional to its degree K. Since edges leading to degree 1 nodes are neither self loops nor double edges, the upper cutoff is at least proportional to the natural cutoff. In reality, the deviation from the exact degree distribution is quite small. Thus, removing the double edges and self loops does not affect the statistical behavior of the tail of the distribution.

Exercises

9.1 Use the generating function method to obtain the equation for the size of the giant component in ER networks (having degree distribution $P(k) = e^{-c}c^k/k!$ where $\langle k \rangle = c$).

9.2 (a) Use the generating function method to obtain the equation for the size of the giant component in random regular networks with all nodes having degree 3.

(b) How will your results change if all degrees were 4? What can you say about a random regular network with all degrees D?

9.3 What is the size of the giant component for a network with half the nodes having degree 3 and half having degree 4? Can you generalize this result?

9.4 What is the size of the giant component for a network with half the nodes having degree 3 and half having degree 1? What is the value of κ for this network?

9.5 Construct the generating function for the component (cluster) sizes for bipartite networks (see Exercise 4.2 for the definition) with degree distributions $P_a(k)$ and $P_b(k)$. See [NSW01] for the solution.

10 Percolation on complex networks

10.1 Introduction

It is well known [BH96, SA94] that in grids and other organized lattices, in any dimension larger than one, a percolation phase transition occurs. The usual percolation model assumes that the sites (nodes) or bonds (links) in the lattice are occupied with some probability (or density), p, and unoccupied with probability $q = 1 - p$. The system is considered percolating if there is a path from one side of the lattice to the other, passing only through occupied links and nodes. When such a path exists, the component or cluster of sites that spans the network from side to side is called the spanning cluster or the infinite cluster The percolation phase transition occurs at some critical density p_c that depends on the type and dimensionality of the lattice.

In networks no notion of side exists. However, as will be seen in this chapter, the ideas of percolation theory can still be applied to obtain useful results. The main difference compared to lattices is that the condition for percolation is no longer the spanning property, but rather the property of having a component (cluster) containing $\mathcal{O}(N)$ nodes, where N is the total original number of nodes in the network. Such a component, if it exists, is termed the **giant component**. The condition of the existence of a giant component above the percolation threshold and its absence below the threshold also applies to lattices, and therefore can be considered as more general than the spanning property.

An interesting property of percolation, called **universality**, is that the behavior at the critical point and near it depends only on the dimensionality of the lattice, and not on the microscopic connection details of the lattice. This behavior is characterized by a set of "critical exponents" that are the same for all d-dimensional lattices, even if they are square, triangular, or hexagonal, site or bond percolation. However, a different set will be obtained for another dimension of a lattice. Furthermore, above some critical dimension ($d_c = 6$ for percolation in d-dimensional lattices), known as the **upper critical dimension**, the critical behavior remains the same, regardless of the dimension. This is due to the insignificance of the loops in high dimensions, which usually easily allows the determination of the critical exponents for high dimensions, using an "infinite-dimensional" or "mean-field" approach. Percolation on ER networks or on Cayley trees has the same critical exponents as for lattices above the upper critical dimension, since no spatial constraints are imposed on these networks. In this chapter we discuss how, in fact, the heterogeneity of the degrees may still affect the critical behavior even above the critical dimension. The heterogeneity of the degrees can be regarded as a breakdown of the translational symmetry that exists in lattices, ER networks, and Cayley trees. In all these cases each node has a typical number of neighbors, whereas in scale-free networks the variation between node degrees is very large. The results are still "mean field" or infinite dimensional in the sense of the insignificance of the loops. However, results that differ from the standard mean-field percolation solution are obtained.

As will be discussed later in this chapter, for scale-free networks with $\gamma > 4$, the critical exponents are the same as for ER graphs and lattices in $d \geq d_c = 6$. In this case, γ is large enough and the translational symmetry exists (at least approximately). For $\gamma < 4$, however, the topology is different, and therefore, the critical exponents are different. Thus, again, scale-free networks can be regarded as a generalization of ER networks.

10.2 Random breakdown

10.2.1 Description

Albert *et al.* [AJB00] suggested modeling the Internet at the router level as a scale-free network (in their model $\gamma \approx 3$). They suggested a scenario in which nodes in the Internet network fail randomly (due to random error or an external cause such as power failures). To model this scenario, they suggested random removal of nodes from the network, after which they calculated the size of the largest remaining cluster. They compared the results of applying this process to a scale-free network to the same model on a random Erdős–Rényi (ER) graph. Interestingly, they found that scale-free networks are much more resilient to this kind of failure than ER graphs. This problem was studied asymptotically for any degree distribution $P(k)$ [CEbH00, CNSW00] and is described in this chapter.

10.2.2 Critical threshold for percolation

The considerations in the previous chapter can be applied to the problem of percolation on a generalized random network with a given degree distribution $P(k)$. If we randomly remove a fraction q of the nodes (or links), the degree distribution of the remaining nodes will change. For instance, nodes with initial degree k_0 will have, after the random removal of nodes, a different number of links, depending on the number of neighbors removed. The new number of links will be binomially distributed. If we begin with a distribution of degrees $P_0(k_0)$, the new distribution of degrees in the network will be:

$$P(k) = \sum_{k_0=k}^{\infty} P_0(k_0) \binom{k_0}{k} p^k (1-p)^{k_0-k}. \tag{10.1}$$

Calculating the first two moments for this distribution, given $\langle k_0 \rangle$ and $\langle k_0^2 \rangle$ for the original distribution, leads to:

$$\langle k \rangle = \sum_{k=0}^{\infty} P(k) k = p \langle k_0 \rangle . \tag{10.2}$$

Similarly, we can calculate:

$$\langle k^2 \rangle = \sum_{k=0}^{\infty} P(k) k^2 = p^2 \langle k_0^2 \rangle + p(1-p) \langle k_0 \rangle . \tag{10.3}$$

Both can be substituted into Eq. (9.2) to find the criterion for criticality. This yields:

$$\kappa \equiv \frac{\langle k^2 \rangle}{\langle k \rangle} = \frac{p^2 \langle k_0^2 \rangle + p(1-p) \langle k_0 \rangle}{p \langle k_0 \rangle} = 2. \tag{10.4}$$

Reorganizing Eq. (10.4), onc obtains the critical threshold for percolation [CEbH00]:

$$p_c = \frac{1}{\kappa_0 - 1}, \tag{10.5}$$

where $\kappa_0 \equiv \langle k_0^2 \rangle / \langle k_0 \rangle$ is calculated using the original distribution, before the nodes are removed.

Equations (9.2) and (10.5) are valid for a wide range of generalized random graphs and distributions. For example, for a Cayley tree – a graph with a fixed degree z and no loops – the criterion from Eq. (10.5) can be used. This yields the critical concentration $p_c = 1 - q_c = 1/(z-1)$, which is well known [BH96]. Another example is a random Erdős–Rényi (ER) graph. In these graphs, edges are distributed randomly and the resulting degree distribution is Poissonian [Bol85]. Applying the criterion from Eq. (9.2) to a Poisson distribution results in:

$$\kappa \equiv \frac{\langle k^2 \rangle}{\langle k \rangle} = \frac{\langle k \rangle^2 + \langle k \rangle}{\langle k \rangle} = 2, \tag{10.6}$$

which reduces to criticality at $\langle k \rangle = 1$, and from Eq. (10.5) follows $p_c = 1/\langle k \rangle$, as known for ER graphs [Bol85].

10.2.3 Scale-free networks

In calculating the threshold for random breakdown, the key parameter, according to (10.5), is the ratio between the second moment and the first moment, κ_0, which we compute by approximating the distribution (2.2) by a continuous distribution.[1] This approximation becomes exact for $1 \ll m \ll K$, and it preserves the essential features of the transition even for small m. Furthermore, if the degree is chosen from a continuous distribution and is then rounded, this approximation is fairly accurate for all values of the cutoff:[2]

$$\kappa_0 = \left(\frac{2-\gamma}{3-\gamma}\right)\frac{K^{3-\gamma} - m^{3-\gamma}}{K^{2-\gamma} - m^{2-\gamma}}. \tag{10.7}$$

When $K \gg m$, this may be approximated as:

$$\kappa_0 \to \left|\frac{2-\gamma}{3-\gamma}\right| \times \begin{cases} m & \text{if } \gamma > 3 \\ m^{\gamma-2}K^{3-\gamma} & \text{if } 2 < \gamma < 3 \\ K & \text{if } 1 < \gamma < 2. \end{cases} \tag{10.8}$$

We see that for $\gamma > 3$ the ratio κ_0 is finite and there is a percolation transition at $p_c \approx \left(\frac{\gamma-2}{\gamma-3}m - 1\right)^{-1}$; for $p < p_c$ the spanning cluster is fragmented and the network is destroyed. However, for $\gamma < 3$ the ratio κ_0 diverges with K and so $p_c \to 0$ when $K \to \infty$ (or $N \to \infty$). The percolation transition does not take place: a spanning cluster exists for arbitrarily large fractions of breakdown, $q < 1$ or $p > 0$. In *finite* systems a transition is always observed, though for $\gamma < 3$ the transition threshold is exceedingly high. For the case of the Internet ($\gamma \approx 5/2$), we have $\kappa_0 \approx K^{1/2} \approx N^{1/3}$. Considering the enormous size of the Internet, $N > 10^6$, one needs to remove randomly about 99% of the nodes before the spanning cluster collapses. As seen in Chapter 9, for $\gamma > 4$ and $m = 1$, calculation of κ shows that it is less than 2 even before the breakdown occurs. This indicates that the network is fragmented and contains no giant component even before breakdown appears. For $m \geq 2$, a spanning cluster exists for every γ.

The size of the spanning cluster can also be measured using the method suggested in [NSW01]. The distribution (10.1) can be substituted into Eq. (9.4) and the other generating functions can be calculated using this distribution, by giving the size of the spanning cluster relative to the undisturbed network. An alternative method [CNSW00] is to build a new generating function, taking the fraction of removed

[1] For a discussion on the discrete distribution, see [DM01a].

[2] Note, however, that if the degree distribution has a cutoff larger than $\sqrt{N}$, the loops become significant for the high-degree nodes, and the approximation becomes inaccurate (see [PSS05] for more details).

nodes into account:

$$G_0(x) = \sum_{k=0}^{\infty} P(k)p(k)x^k, \tag{10.9}$$

where $p(k) = 1 - q(k)$ is the probability that a node of degree k is not removed. This equation replaces Eq. (9.4). For a random breakdown $p(k) = p = 1 - q$ is independent of k. The size of the infinite cluster is then given by

$$P_\infty = G_0(1) - G_0(u), \tag{10.10}$$

where u is the smallest positive solution of

$$u = 1 - G_1(1) + G_1(u). \tag{10.11}$$

The analytical and numerical results for a random breakdown (for simulation details, see Appendix C) can be seen in Figures 10.1 and 10.2.

10.3 Intentional attack

10.3.1 Description

Another model, originally suggested in [AJB00], is that of intentional attack on the most highly connected nodes in the network. In this model an attacker (for example, a computer hacker trying to cause maximum damage to the network) initiates by some means an intentional attack on the most highly connected nodes in the network, causing breakdown of those nodes.

In a homogeneous network, such as an ER network (and certainly in random regular networks where all degrees are equal) the high-degree nodes are not very different in degree compared with any randomly chosen node. Therefore, it is expected that this kind of attack will not be qualitatively different from random removal. This is seen numerically in [AJB00, AJB01], where, although the threshold is different, the behavior of the transition in general is similar to that of the random removal percolation transition. In a scale-free network, however, an attack such as this is expected to cause more extensive damage than a random attack, and as will be shown in the next sections, can cause even networks robust to random breakdown to collapse.

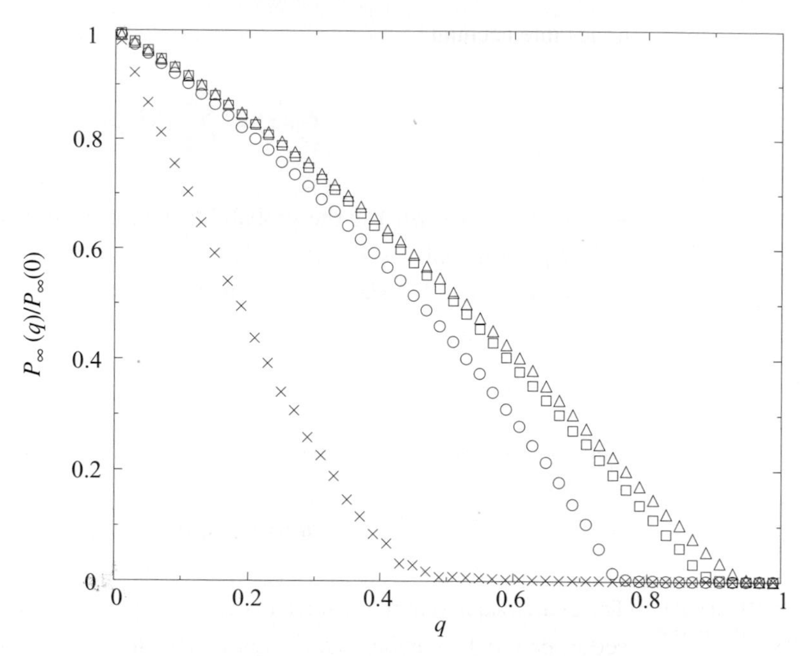

Figure 10.1 Percolation transition for networks with power-law degree distribution. Plotted is the fraction of nodes that remain in the spanning cluster after breakdown of a fraction q of all nodes, $P_\infty(q)/P_\infty(0)$, as a function of q, for $\gamma = 3.5$ (crosses) and $\gamma = 2.5$ (other symbols), as obtained from computer simulations of up to $N = 10^6$. In the $\gamma = 3.5$ case, it can be seen that for $p > p_c \approx 0.5$ the spanning cluster disintegrates and the network becomes fragmented. However, for $\gamma = 2.5$ (the case of the Internet), the spanning cluster persists up to nearly 100% breakdown. The different curves for $N = 100$ (circles), $N = 10^3$ (squares), and $N = 10^4$ (triangles) illustrate the finite size effect: the transition exists only for finite networks, whereas the critical threshold q_c approaches 1 as the networks grow in size.

10.3.2 Theory

Consider now intentional attack, or sabotage [AJB00], whereby a fraction q of the nodes with the highest degree is removed. (The links emanating from the nodes are removed as well.) This has the following effect: (a) the cutoff degree K is reduced to some new value, $\tilde{K} < K$, and (b) the degree distribution of the remaining nodes is no longer scale free, but is changed, because of the removal of many of their links. The upper cutoff K before the attack may be estimated, as in Eq. (2.4)

$$\sum_{k=K}^{\infty} P(k) = \frac{1}{N}, \tag{10.12}$$

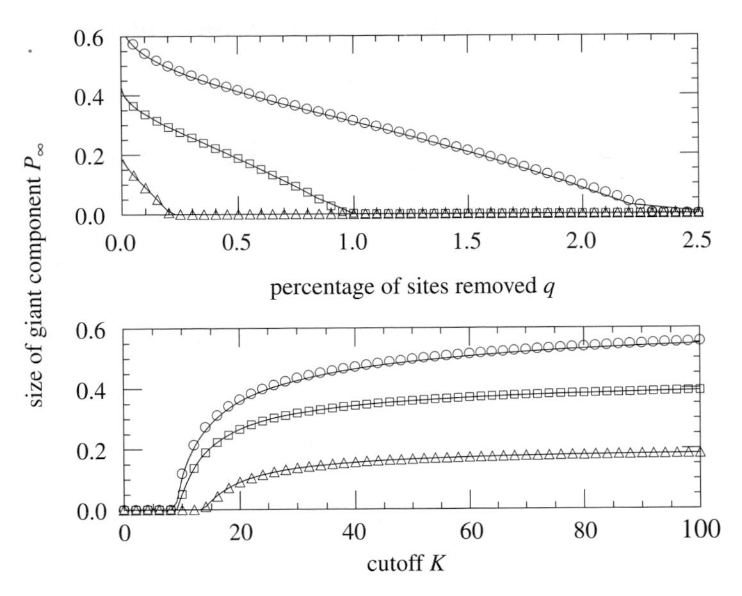

Figure 10.2

Size of the giant component P_∞ in networks with power-law degree distribution and all nodes with degree greater than K unoccupied, for $\gamma = 2.4$ (circles), 2.7 (squares), and 3.0 (triangles). Points are simulation results for systems with 10^7 vertices; solid lines are the exact solution. Upper frame: as a function of fraction of vertices unoccupied. Lower frame: as a function of the cutoff parameter K. After [CNSW00].

where N is the total number of nodes in the original network. Similarly, the new cutoff $\tilde{K}$, after the attack, can be estimated from

$$\sum_{k=\tilde{K}}^{K} P(k) = \sum_{k=\tilde{K}}^{\infty} P(k) - \frac{1}{N} = q. \tag{10.13}$$

If the size of the system is large, $N \gg 1/q$, the original cutoff K may be safely ignored. We can then obtain $\tilde{K}$ approximately by replacing the sum with an integral:

$$\tilde{K} = mq^{1/(1-\gamma)}. \tag{10.14}$$

Next, we estimate the impact of the attack on the distribution of the remaining nodes. The removal of a fraction $q = 1 - p$ of the nodes with the highest degree results in a random removal of links from the remaining nodes – links that had connected the removed nodes with the remaining nodes. The probability $\tilde{q}$ of a link leading to a deleted node equals the ratio of the number of links belonging to deleted nodes to the total number of links:

$$\tilde{q} = \sum_{k=\tilde{K}}^{K} \frac{kP(k)}{\langle k_0 \rangle}, \tag{10.15}$$

where $\langle k_0 \rangle$ is the initial average degree. With the usual continuous approximation, and neglecting K, this yields

$$\tilde{q} = \left(\frac{\tilde{K}}{m}\right)^{2-\gamma} = q^{(2-\gamma)/(1-\gamma)}, \tag{10.16}$$

for $\gamma > 2$. For $\gamma = 2$, $\tilde{q} \to 1$ or $\tilde{p} \to 0$, since just a few nodes of very high degree control the entire connectedness of the system. Indeed, consider a finite system of N nodes and $\gamma = 2$. The upper cutoff $K \approx N$ must then be taken into account, and approximating Eq. (10.15) by an integral yields $\tilde{q} = \ln(Nq/m)$. That is, for $\gamma = 2$, very small values of q are needed to destroy an arbitrarily large fraction of the links as $N \to \infty$.

Using these results, we can compute the effect of intentional attack, using the theory described in Section 10.2 for random removal of nodes [CEbH00]. Essentially, the network after attack is equivalent to a scale-free network with a new cutoff $\tilde{K}$, which has undergone random removal of a fraction $\tilde{q}$ of its nodes. This can be understood as the result of two processes. (a) Removal of the highest degree nodes reduces the upper cutoff. Since this effect changes the degree distribution, κ_0 needs to be recalculated accordingly. (b) Removal of the links leading to the removed nodes. The probability of removing a link is $\tilde{q}$ – the probability that a randomly chosen link leads to one of the removed nodes – and all links have the same probability of being deleted. Since this effect has an influence on the probability distribution described in Eq. (10.1), the result in Eq. (10.5) can be used, with $\tilde{q}$ replacing q. Note that for random node deletion the probability of a link leading to a deleted node is identical to the fraction of deleted nodes.

Although the number of nodes removed in an intentional attack differs from the random breakdown analogy presented above (i.e., reducing the cutoff and then randomly removing a fraction $\tilde{q}$ of the links), this affects the size of the spanning cluster (see below) but not the criterion for the critical point. This is because the transition point is defined as the point where the spanning cluster starts transforming between zero fraction and a finite fraction of the whole network. A finite fraction of the remaining nodes is also a finite fraction of the original network, and similarly, zero fraction of the original network is a zero fraction of the remaining network. Thus, the difference has no effect on p_c.

We therefore use Eqs. (10.5) and (10.7), but with $\tilde{q} = (\tilde{K}/m)^{2-\gamma}$ replacing q_c and $\tilde{K}$ replacing K. This yields

$$(\tilde{K}/m)^{2-\gamma} - 2 = \frac{2-\gamma}{3-\gamma} m[(\tilde{K}/m)^{3-\gamma} - 1], \tag{10.17}$$

which can be solved numerically to obtain $\tilde{K}(m, \gamma)$, and then $q_c(m, \gamma)$ can be derived from Eq. (10.14). In Figure 10.3 we plot p_c – the critical fraction of nodes which must

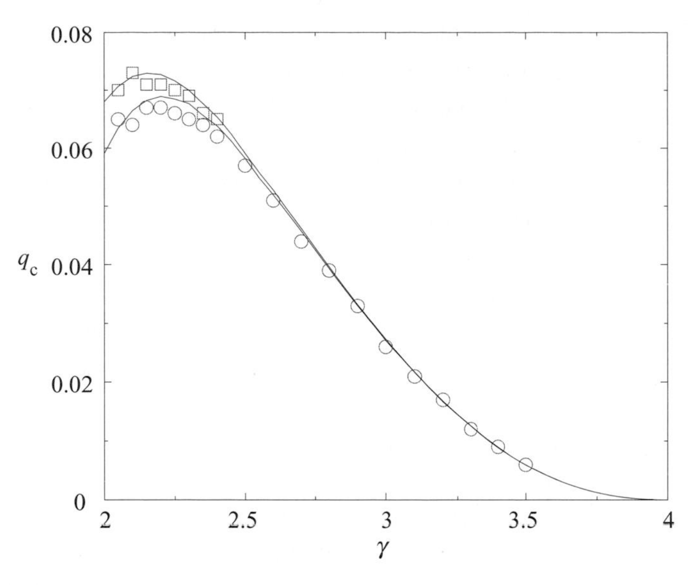

Figure 10.3 Critical probability, q_c, as a function of γ, for networks of size $N = 500\,000$ (circles) and $N = 64\,000$ (squares) and $m = 1$. Lines represent the analytical solution, obtained from Eqs. (10.13) and (10.15). After [CEbH01].

be removed in the intentional attack strategy to disrupt the network – computed in this fashion, and compared to results from numerical simulations. A phase transition exists (at a finite q_c) for all $\gamma > 2$ values. The decline in q_c for large γ values is explained by the fact that as γ increases, the spanning cluster becomes smaller in size, even before an attack. (Furthermore, for $m < 2$, the original network is disconnected for some large enough γ.) The decline in q_c as $\gamma \to 2$ results from the critically high degree of just a few nodes: their removal disrupts the whole network. This was already argued in [AJB00]. We noted that for infinite systems $p_c \to 0$ as $\gamma \to 2$. The critical fraction q_c is rather sensitive to the lower degree cutoff m. For larger m (the case of $m = 1$ is shown in Figure 10.3), the networks are more robust, though they still undergo a transition at a finite $q_c < 1$.

10.4 Critical exponents

10.4.1 Introduction

One of the most well studied properties of critical phenomena is the behavior of the transition at and near criticality. Usually, phase transitions are thought to be controlled by an external parameter, such as the temperature in most physical systems, or the occupation density, p, in the percolation case. When the parameter reaches some critical value, the system undergoes the transition, usually manifested by a discontinuous jump in some order parameter (such as the volume density or magnetization, or, in our case, the relative size of the largest component, P_∞) or one of its derivatives. The scaling of the order parameter with the system size at the transition point, as well as the laws governing its behavior at the neighborhood of the critical point are usually universal properties, depending only on the spatial dimension of the system, as well as the nature and symmetries of the order parameter. Other universal properties are the divergence of sizes such as the correlation length, representing the typical geometrical distance of points for which the value of the order parameter is correlated, the average size of a finite component in a percolating network, and the susceptibility of a magnetic system.

In the case of the percolation phase transition in ER networks, the critical exponents coincide with the critical exponents of high ($d > 6$) and infinite dimensional models. This is also true for scale-free networks with $\gamma > 4$. However, for $\gamma < 4$ the critical exponents are different. As in most critical systems, these exponents do not depend on the microscopic parameters of the network, such as m and $\langle k \rangle$. They do depend, however, on the behavior of the tail of the distribution, determined by γ, which represents the level of inhomogeneity in the network.

10.4.2 Size of the giant component at criticality

In Section 9.2.2 the generating functions $G_0(x)$ and $G_1(x)$ were built for the degree distribution (Eqs. (9.4) and (9.5)). Let $H_1(x)$ be the generating function for the probability of reaching a branch of a given size by following a link. After dilution of a fraction q of the nodes (the remaining concentration is $p = 1 - q$), $H_1(x)$ satisfies the self-consistent equation

$$H_1(x) = 1 - p + pxG_1(H_1(x)) . \tag{10.18}$$

Since $G_0(x)$ is the generating function for the degree of a node, the generating function for the probability of a node to belong to an n-node cluster is

$$H_0(x) = 1 - p + pxG_0(H_1(x)) . \tag{10.19}$$

$H_0(1)$ is the probability that a node belongs to a cluster of any *finite* size. Thus, below the percolation transition $H_0(1) = 1$, while above the transition, there is a finite probability that a node belongs to the infinite spanning cluster: $P_\infty = 1 - H_0(1)$ (see

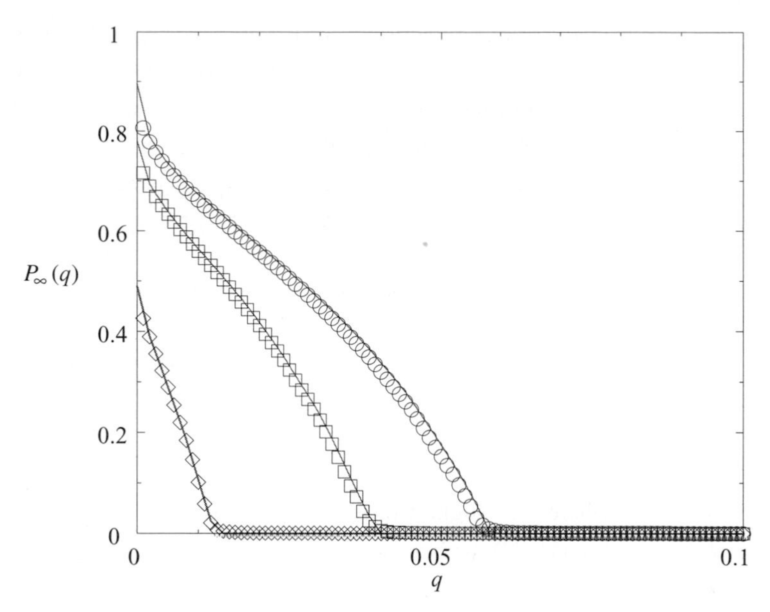

Figure 10.4 Fraction of nodes belonging to the spanning cluster, P_∞, as a function of the fraction of removed nodes, q, for networks with $\gamma = 2.5$ (circles), $\gamma = 2.8$ (squares), and $\gamma = 3.3$ (diamonds). Lines represent the analytical result from Eqs. (9.8) and (9.9). Both the simulation and analysis are for system size $N = 500\,000$, and $m = 1$. After [CEbH01].

Figure 10.4). It follows that

$$P_\infty(p) = p\left(1 - \sum_{k=0}^{\infty} P(k)u^k\right), \tag{10.20}$$

where $u \equiv H_1(1)$ is the smallest positive root of

$$u = 1 - p + \frac{p}{\langle k \rangle}\sum_{k=0}^{\infty} kP(k)u^{k-1}. \tag{10.21}$$

This equation can be solved numerically and the solution can be substituted into Eq. (10.20), yielding the size of the spanning cluster in a network of arbitrary degree distribution, at concentration p [CNSW00].

We now compute the order parameter critical exponent β. Near criticality, the probability of belonging to the spanning cluster or, alternatively, the relative size of the spanning cluster, behaves as $P_\infty \sim (p - p_c)^\beta$. For infinite-dimensional systems (such as a Cayley tree or an ER network), it is known that $\beta = 1$ [BH96, bH00, SA94]. This is regarded as the standard mean-field result, where the constraints imposed by the lattice become irrelevant. Indeed, for percolation in d-dimensional lattices with $d \geq d_c = 6$, the standard mean-field results are valid. This standard mean-field approximation is not always valid, however, for scale-free networks.

To study the critical behavior, we begin by analyzing the singularities of Eqs. (10.20) and (10.21). Equation (10.20) has no special behavior at $p = p_c$; the singular behavior comes from u. Also, criticality $P_\infty = 0$ and Eq. (10.20) imply that $u = 1$. We therefore examine Eq. (10.21) for $u = 1 - \epsilon$ and $p = p_c + \delta$:

$$1 - \epsilon = 1 - p_c - \delta + \frac{(p_c + \delta)}{\langle k \rangle} \sum_{k=0}^{\infty} k P(k)(1 - \epsilon)^{k-1}. \tag{10.22}$$

The sum in (10.22) has the asymptotic form

$$\sum_{k=0}^{\infty} k P(k) u^{k-1} \sim \langle k \rangle - \langle k(k-1) \rangle \, \epsilon + \frac{1}{2} \langle k(k-1)(k-2) \rangle \, \epsilon^2 + \cdots + c\Gamma(2-\gamma)\epsilon^{\gamma-2}, \tag{10.23}$$

where the highest-order analytic term is $\mathcal{O}(\epsilon^n)$, $n = \lfloor \gamma - 2 \rfloor$. Using this in Eq. (10.22), with $q_c = 1/(\kappa - 1) = \langle k \rangle / \langle k(k-1) \rangle$, we obtain

$$\frac{\langle k(k-1) \rangle^2}{\langle k \rangle} \delta = \frac{1}{2} \langle k(k-1)(k-2) \rangle \, \epsilon + \cdots + c\Gamma(2-\gamma)\epsilon^{\gamma-3}. \tag{10.24}$$

The divergence of δ as $\gamma < 3$ confirms the lack of a phase transition in that regime. Thus, limiting ourselves to $\gamma > 3$, and keeping only the dominant term as $\epsilon \to 0$, Eq. (10.24) implies

$$\epsilon \sim \begin{cases} \left(\dfrac{\langle k(k-1) \rangle^2}{c \langle k \rangle \Gamma(2-\gamma)} \right)^{\frac{1}{\gamma-3}} \delta^{\frac{1}{\gamma-3}} & 3 < \gamma < 4 \\ \dfrac{2 \langle k(k-1) \rangle^2}{\langle k \rangle \langle k(k-1)(k-2) \rangle} \delta & \gamma > 4. \end{cases} \tag{10.25}$$

Returning to P_∞, Eq. (10.20), we see that the singular contribution in ϵ is dominant only for the irrelevant range of $\gamma < 2$. For $\gamma > 3$, we find $P_\infty \sim p_c \langle k \rangle \epsilon \sim (p - p_c)^\beta$. Comparing this to (10.25), we finally obtain

$$\beta = \begin{cases} \frac{1}{\gamma - 3} & 3 < \gamma < 4 \\ 1 & \gamma > 4. \end{cases} \tag{10.26}$$

We see that the order parameter exponent β attains its standard mean-field value only for $\gamma > 4$. For $\gamma < 4$, β is modified and becomes larger than 1. Thus, the derivative of the order parameter, P_∞, with respect to p is continuously zero at the phase transition point. The *order* of the phase transition is determined by the discontinuity in the order parameter and its derivative. If the order parameter is discontinuous at the transition point, the transition is called first order; if the order parameter is continuous but its derivative is discontinuous, it is considered a second-order phase transition. Discontinuity in the second derivative determines a third-order phase transition, etc. Thus, the percolation transition for $\gamma < 4$ is higher than the second order: for $3 + \frac{1}{n-1} < \gamma < 3 + \frac{1}{n-2}$ the transition is of the nth order. The result (10.26) was reported before in [CEbH01], and was also found independently in a different but related model of virus spreading [MPV02, PV01b]. The existence of an infinite-order phase transition at $\gamma = 3$ for growing networks of the Barabási–Albert model was reported elsewhere [CHK+01, DMS01a]. These examples suggest that the critical exponents are not model dependent but depend only on γ.

For networks with $\gamma < 3$, the transition still exists, though at a vanishing threshold, $p_c = 0$. The sum in Eq. (10.22) becomes:

$$\sum_{k=0}^{\infty} kP(k)u^{k-1} \sim \langle k \rangle + c\Gamma(2-\gamma)\epsilon^{\gamma-2} . \tag{10.27}$$

Using this in conjunction with Eq. (10.21), and remembering that here $p_c = 0$ and therefore $p = \delta$, leads to

$$\epsilon = \left(\frac{-c\Gamma(2-\gamma)}{\langle k \rangle} \right)^{\frac{1}{3-\gamma}} \delta^{\frac{1}{3-\gamma}} , \tag{10.28}$$

which implies

$$\beta = \frac{1}{3-\gamma}, \qquad 2 < \gamma < 3 . \tag{10.29}$$

In other words, the transition in $2 < \gamma < 3$ is a mirror image of the transition in $3 < \gamma < 4$. An important difference is that $p_c = 0$ is not γ-dependent for $2 < \gamma < 3$, and the prefactor of P_∞ diverges as $\gamma \to 2$ but remains finite as $\gamma \to 4$, implying that for $\gamma \to 2$ the giant component undergoes a discontinuous jump at the transition point.

10.4.3 Finite component size distribution

It is known that for percolation in any dimension [BH96, SA94] and for random graphs of arbitrary degree distribution [NSW01], the finite components near criticality follow the scaling form

$$n_s \sim s^{-\tau} e^{-s/s^*} . \tag{10.30}$$

Here s is the component size, and n_s is the number of components of size s. At criticality $s^* \sim |p - p_c|^{-\sigma}$ diverges and the tail of the distribution behaves as a power law.

We now derive the exponent τ for scale-free networks. The probability that a node belongs to an s-component is $p_s = sn_s \sim s^{1-\tau}$, and is generated by H_0:

$$H_0(x) \equiv \sum p_s x^s \ . \tag{10.31}$$

The singular behavior of $H_0(x)$ stems from $H_1(x)$, as can be seen from Eq. (10.19). $H_1(x)$ itself can be expanded from Eq. (10.18), by using the asymptotic form (10.23) of G_1. We let $x = 1 - \epsilon$, as before, but work at the critical point, $p = p_c$. With the notation $\phi(\epsilon) = 1 - H_1(1 - \epsilon)$, we finally get (note that at criticality $H_1(1) = 1$):

$$-\phi = -p_c + (1-\epsilon)p_c \left[1 - \frac{\phi}{p_c} + \frac{\langle k(k-1)(k-2)\rangle}{2\langle k\rangle}\phi^2 + \cdots + c\frac{\Gamma(2-\gamma)}{\langle k\rangle}\phi^{\gamma-2} \right]. \tag{10.32}$$

From this relation we extract the singular behavior of H_0: $\phi \sim \epsilon^y$. Then, using Tauberian theorems [Wei94] it follows that $p_s \sim s^{-1-y}$, hence $\tau = 2 + y$.

For $\gamma > 4$, the term proportional to $\phi^{\gamma-2}$ in (10.32) can be neglected. The linear term $\epsilon\phi$ can be neglected as well, due to the factor ϵ. This leads to $\phi \sim \epsilon^{1/2}$ and to the standard mean-field result [BH96]

$$\tau = \frac{5}{2}, \qquad \gamma > 4 \ . \tag{10.33}$$

For $\gamma < 4$, the terms proportional to $\epsilon\phi$, ϕ^2 can be neglected, leading to $\phi \sim \epsilon^{1/(\gamma-2)}$ and

$$\tau = 2 + \frac{1}{\gamma - 2} = \frac{2\gamma - 3}{\gamma - 2}, \qquad 2 < \gamma < 4 \ . \tag{10.34}$$

Note that for $2 < \gamma < 3$ the percolation threshold is strictly at $p_c = 0$. Here we analyze Eq. (10.32) at $p = \delta$ small but fixed, taking the limit $\delta \to 0$ at the very end. The case $2 < \gamma < 3$, τ in Eq. (10.34) represents the singularity of the distribution of branch sizes. For the distribution of component sizes in this range, one has to consider the singularity of x in Eq. (10.19) leading to $\tau = 3$ for this range.

For the growing networks model of Barabási–Albert with $\gamma = 3$, it has been shown that $sn_s \propto (s \ln s)^{-2}$ [DMS01a]. This is consistent with $\tau = 3$ plus a logarithmic correction. Related results for scale-free trees have been presented in [BCK01].

At the transition point the largest component, S can be obtained from the finite component distribution by assuming the integral over the tail of the distribution to be equal to $1/N$. This results in

$$S \propto N^{\tau-1} = N^{(\gamma-2)/(\gamma-1)}. \tag{10.35}$$

For $\gamma = 4$ this reduces to the known $N^{2/3}$ found for ER networks [Bol85]. For $\gamma \to 3$, $S \propto N^{1/2}$. It is not yet clear whether the result (10.35) also has a meaningful interpretation for $\gamma < 3$.

10.4.4 Finite component size cutoff

The critical exponent σ, for the cutoff component size, can also be derived directly. Finite-size scaling arguments predict [SA94] that

$$p_c(\infty) - p_c(N) \sim N^{-\frac{1}{d\nu}} = N^{-\frac{\sigma}{\tau-1}}, \tag{10.36}$$

where N is the number of nodes in the network, ν is the correlation length critical exponent, $\xi \sim (p - p_c)^{-\nu}$, and d is the dimensionality of the embedding space. Using a continuous approximation of the distribution (2.2), one obtains [CEbH00]

$$\kappa \approx \left(\frac{2-\gamma}{3-\gamma}\right)\frac{K^{3-\gamma} - m^{3-\gamma}}{K^{2-\gamma} - m^{2-\gamma}}, \tag{10.37}$$

where $K \sim N^{1/(\gamma-1)}$ is the largest node degree of the network. For $3 < \gamma < 4$, this and Eq. (10.5) yield

$$p_c(\infty) - p_c(N) \sim \Delta\kappa \sim K^{3-\gamma} \sim N^{\frac{3-\gamma}{\gamma-1}}, \tag{10.38}$$

which in conjunction with Eq. (10.36) leads to

$$\sigma = \frac{\gamma-3}{\gamma-2}, \qquad 3 < \gamma < 4\,. \tag{10.39}$$

For $\gamma > 4$ we recover the standard mean-field result $\sigma = 1/2$. Note that Eqs. (10.36), (10.26), and (10.34) are consistent with the known scaling relation: $\sigma\beta = \tau - 2$ [BH96, bH00, SA94]. For $2 < \gamma < 3$, $q_c(\infty) = 0$ and $q_c(N) \sim K^{\gamma-3} \sim N^{(\gamma-3)/(\gamma-1)}$. Therefore,

$$\sigma = \frac{3-\gamma}{\gamma-2}, \qquad 2 < \gamma < 3, \tag{10.40}$$

which is again consistent with the scaling relation $\sigma\beta = \tau - 2$ (cf. Eq. (10.29)).

10.4.5 Fractal and upper critical dimensions

As seen in Chapter 6, the distances between nodes in scale-free networks, and generalized random graphs in general are very small, of order $\log N$ or even $\log\log N$. However, the diluted case is essentially the same as infinite-dimensional percolation. In this case, there is no notion of geometrical distance (since the graph is not embedded in a Euclidean space), but only of a distance along the graph (which is the shortest distance along bonds). It is known from infinite-dimensional percolation theory that the fractal dimension at criticality is $d_l = 2$ [BH96]. Therefore, the average (chemical) distance l between pairs of nodes on the largest component at criticality

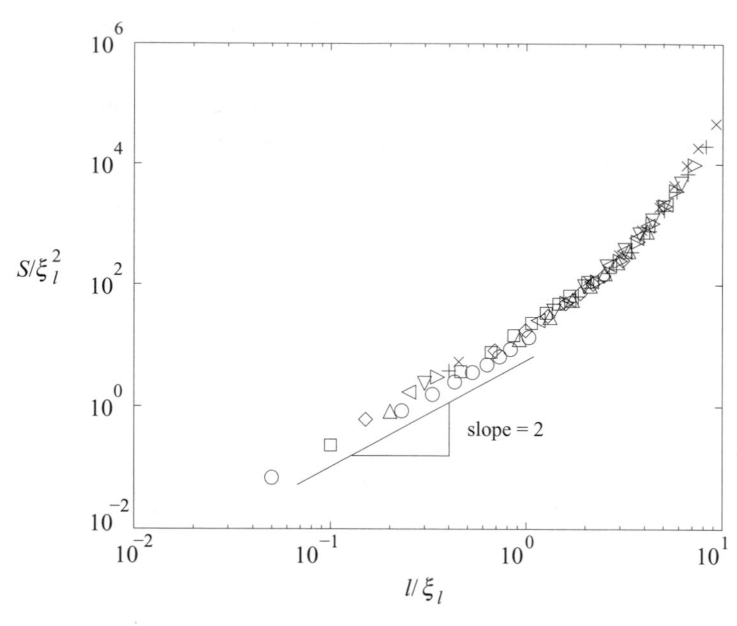

Figure 10.5 Mass (number of nodes), S, as a function of distance, l, on the spanning component for an ER network at criticality. The correlation length is $\xi_l = |p - p_c|^{-1}$. Note that for $l/\xi_l < 1$, the slope is 2, corresponding to the behavior in the critical regime, whereas for $l/\xi_l > 1$, M grows exponentially with l, corresponding to the well-connected regime. After [CEbH01].

behaves as

$$l \sim \sqrt{S}, \tag{10.41}$$

where S is the number of nodes in the largest component. Figure 10.5 shows simulation results supporting (10.41). This is analogous to percolation in finite dimensions, where in length-scales smaller than the correlation length the component is a fractal with dimension d_f and above the correlation length the component is homogeneous and has the dimension of the embedding space. In our infinite-dimensional case, the crossover between these two behaviors occurs around the correlation length $\xi_l \approx |p_c - p|^{-\nu_l}$, with $\nu_l = 1$. We denote by ξ_l and ν_l the correlation length and the exponent, respectively, in the shortest path metric, l. Since for infinite dimensions (of an embedding space) $\nu = \frac{1}{2}$ and $\ell \sim r^2$, it follows that $\nu_l = 1$.

For scale-free networks, the result for ν_l is the same, but d_l is different. Below the transition all components are finite and almost all of them are trees. The correlation length can be defined using the formula [BH96]:

$$\xi_l^2 = \frac{\sum l^2 g(l)}{\sum g(l)}, \tag{10.42}$$

where $g(l)$ is the number of nodes at a distance l from a random node (and belonging to the same component). The number of nodes in the l shell is approximately $\langle k \rangle (\kappa - 1)^{l-1}$ [NSW01]. Since $\kappa - 1 = (\kappa_0 - 1)q$ and $p_c = 1/(\kappa_0 - 1)$, we get $g(l) = c(1 - \delta)^l$, where $\delta = p - p_c$. This leads to $\xi_l \sim (p - p_c)^{-1}$, i.e., $\nu_l = 1$. Above the threshold, the finite components can be seen as a random graph with the residual degree distribution of nodes not included in the gaint component [MR98]. That is, the degree distribution for nodes in the finite components is

$$P_r(k) = P(k)u^k, \tag{10.43}$$

where u is the solution of Eq. (10.21). Using this distribution, we can define κ_r for the finite components. This adds a term proportional to $\epsilon^{\gamma-3}$ to the expansion of ξ_l. But, since $\delta \propto \epsilon^{\gamma-3}$ (10.25), this leads again to $\nu_l = 1$.

Using ν_l, we can find the fractal dimension of the network at criticality. The chemical dimension $d_l = 1/\sigma\nu_l$. Therefore,

$$d_l = \frac{\gamma - 2}{\gamma - 3} . \tag{10.44}$$

Any path on the network, when embedded in a Euclidean space above the upper critical dimension, can be regarded as a random walk where $\nu = \nu_l/2$ [BH96]. Therefore, the fractal dimension of the component is

$$d_f = \frac{1}{\nu\sigma} = 2\frac{\gamma - 2}{\gamma - 3}. \tag{10.45}$$

As seen above, the size of the largest component at criticality scales as $S = N^{1/(\tau-1)}$. Assuming the network is embedded in a Euclidean space with some dimension d_c, and linear size L, the total network size is $N = L^{d_c}$, and the largest component has a fractal dimension d_f, and therefore its size is $S \sim L^{d_f}$. This leads to the relation $d_c = d_f(\tau - 1)$. Therefore, the upper critical dimension of the embedding space is,

$$d_c = \frac{\tau - 1}{\nu\sigma} = 2\frac{\gamma - 1}{\gamma - 3}. \tag{10.46}$$

These dimensions reduce to the standard mean-field values 2, 4, and 6 [BH96], respectively, for $\gamma = 4$. These results also hold for $\gamma > 4$ and for ER networks. Equations (10.44)–(10.46) show anomalous mean-field values for $3 < \gamma < 4$, owing to the heterogeneity in the degrees of the nodes in the network, which breaks the translational symmetry. A different method to obtain similar results for the fractal dimensions can be found in Burda *et al.* [BCK01].

A direct method for calculating the chemical dimension, d_l, is also possible. Denoting the generating function of the number of nodes on the lth layer of some branch, as $N_l(x)$, we get

$$N_{l+1}(x) = G_1(N_l(x)) . \tag{10.47}$$

We are interested in the behavior of the average number of nodes at a chemical distance l for those branches that have at least l layers. Since we expand exactly at criticality, the average branching factor is exactly 1, and therefore $N_l(1) = 1$ for any l. Hence, A_l, the average number of nodes for surviving branches, is

$$A_l = \frac{1}{1 - N_l(0)}, \tag{10.48}$$

since $N_l(0)$ is the probability that the branching process dies out before the lth layer. At criticality the branching process will die out with probability $N_l(0) \to 1$ as $l \to \infty$, and therefore for large l we can take $N_l(0) = 1 + \epsilon_l$. Expanding G_1 at criticality, one obtains (Eqs. (10.22) and (10.23), with $\delta = 0$)

$$G_1(1 - \epsilon) = 1 - \epsilon + \frac{c\Gamma(2-\gamma)}{\langle k^2 \rangle - \langle k \rangle}\epsilon^{\gamma-2} + \cdots . \tag{10.49}$$

Substituting $N_l(0) = 1 - \epsilon_l$ into Eq. (10.47), one obtains

$$1 - \epsilon_{l+1} = 1 - \epsilon_l - \frac{c\Gamma(2-\gamma)}{\langle k^2 \rangle - \langle k \rangle}\epsilon_l^{\gamma-2} + \cdots . \tag{10.50}$$

Guessing a solution of the form $\epsilon_l \approx Bl^{-x}$, we obtain

$$B(l+1)^{-x} \approx B(l^{-x} - xl^{-x-1}) = Bl^{-x} - \frac{c\Gamma(2-\gamma)}{\langle k^2 \rangle - \langle k \rangle}(Bl^{-x})^{\gamma-2}, \tag{10.51}$$

implying that $x = 1/(\gamma - 3)$, and $N_l(0) \sim l^x$. Noting that the mass of the branch is the sum of the layers up to the lth layer, we get $d_l = x + 1 = (\gamma - 2)/(\gamma - 3)$, the same as in Eq. (10.44). The two methods therefore produce consistent results.

The behavior of several thermodynamical models on random networks has been studied in the past; in particular, spin interaction models have been studied in [KS87]. The critical exponents for these cases are also known to be the standard mean-field exponents [BH96]. For scale-free networks, the critical exponents for the Ising model [DGM02b, LVVZ02] and for general Landau theory [GDM02] have been calculated, and were shown to be anomalous even for $4 < \gamma < 5$. More details will be given in Chapter 16.

10.5 Percolation in networks with correlations

When degree-degree correlations are present in a network, the above considerations no longer hold. The degree distribution and expected degree for a node reached by following a link is no longer independent of the node through which we arrived at the link. Therefore, the criterion using κ is no longer valid, and should be replaced with a more general criterion. We now turn to finding this criterion for a network with degree-degree correlation where a fraction q of the nodes has been randomly removed (i.e., a fraction $p = 1 - q$ of the nodes remain).

To find the threshold for percolation in correlated networks, we use again the fact that locally the network is tree-like and that below the percolation threshold almost all components are trees. Assume that the probability of a link emanating from a node of degree k to lead to a node of degree k' is $P(k'|k)$. Starting from a random node, we start exploring the network one layer at a time. Suppose at the lth layer there are $n_l(k)$ nodes of degree k for each k. Each such node has $k - 1$ links going to the $l + 1$ layer. The probability of each such link to lead to an undeleted node is p, and the probability of each of these nodes to have degree k' is $P(k'|k)$. Therefore, in the $l + 1$ layer, the number of nodes of each degree, k' will be:

$$n_{l+1}(k') = \sum_k (k-1)n(k)pP(k'|k)\,. \tag{10.52}$$

This equation determines the number of nodes of each degree in each layer given the previous one. Note that if we consider $n_l(k)$ to be a vector, where k is the index, the equation can be written as a matrix equation,

$$\vec{n}_{l+1} = p\mathbf{A}\vec{n}_l, \tag{10.53}$$

where the elements of the matrix $\mathbf{A}$ are

$$A_{k',k} = (k-1)n(k)P(k'|k)\,. \tag{10.54}$$

In Appendix A we discuss such processes. Since all the matrix elements are non-negative, this matrix has one largest eigenvalue with an appropriate non-negative eigenvector. This largest eigenvalue determines the percolation threshold of the network. If this value is above p^{-1} (i.e., the largest eigenvalue of $q\mathbf{A}$ is larger than 1), then the process will continue indefinitely (until $\mathcal{O}(N)$ nodes are reached) and the network is percolating. If the largest eigenvalue is below p^{-1}, no giant component exists. The criterion is therefore

$$1 - q_c = p_c = \lambda_{max}^{-1}, \tag{10.55}$$

where λ_{max} is the largest eigenvalue of the matrix $\mathbf{A}$ defined in Eq. (10.54).

10.5.1 Percolation in correlated scale-free networks

As shown above, in scale-free networks with $\gamma < 3$, the percolation threshold is $p_c \to 0$. However, this was shown above for uncorrelated networks. The question arising is whether this also holds true for networks with degree-degree correlations.

This question was answered in [BPV03].[3] To prove that, one may consider the matrix $A^2 = \sum_{k''} A(k, k'')A(k'', k')$. Consider now the product $\vec{v}^t A^2 \vec{v}$, where $\vec{v}$ is the vector whose elements are $v_k = kP(k)$. For $\gamma > 2$ the vector is normalizable, i.e., $\sum_k (kP(k))^2 < \infty$. Using the equality $kP(k'|k)P(k) = k'P(k|k')P(k')$, stemming from the fact that the number of links going from a node of degree k to a node of degree k' equals the number going the opposite way, one can conclude that the product $\vec{v}^t A^2 \vec{v}$ diverges for $\gamma < 3$. This implies that the largest eigenvalue of A^2 must diverge and therefore also the largest eigenvalue of A diverges, proving that the percolation threshold is 0. For further results on percolation in correlated networks, see also [GDM08].

10.6 k-core percolation: fault tolerant networks

As a final topic in this chapter, we will discuss the notion of k-core percolation. This is a generalization of the regular percolation model, leading to an interesting phase transition.

The k-core of a network is defined as the largest subgraph (i.e., component if it is connected) consisting only of nodes of degree k and above. This core can be obtained from the original network by removing all nodes with degree below k and repeating this process again and again (each time removing every node whose degree dropped below k after the previous round) until no more nodes can be removed. At this stage we are left with a network consisting only of nodes of degree k and above connected among them.[4] See Figure 10.6 for an illustration.

The main interest in the structure of the k-core stems from its fault tolerance. For ER networks, it was proven that the k-core is k-connected. This means that between any two nodes in the k-core there are k independent paths (i.e., not sharing any nodes or links). This is desirable in many cases in order to deal with malfunctions in the network and with congestion. In the k-core, it is guaranteed that less than k failures *anywhere* will not affect the ability of any pair of nodes to communicate. Similarly, any $k - 1$ congested links or nodes can always be bypassed for communication between any pair of nodes.

To prevent confusion between the degree and the core degree k, we will use a C for the core minimum degree and a k for the node degrees.

[3] In fact, the proof in [BPV03] is for the epidemic threshold, discussed in Chapter 14. However, here we adapt the argument to discuss percolation.

[4] There may be more than one component left at the end of the process. Our main interest is in the largest one. For $k \geq 3$ and random graphs the probability of having more than one component in the k-core is negligible.

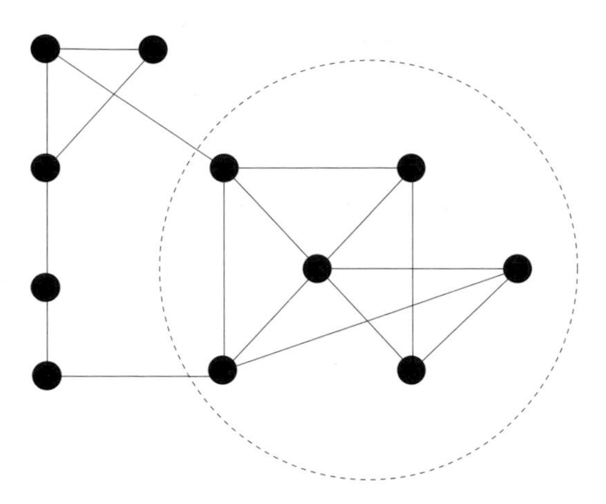

Figure 10.6 An illustration of a network with the 3-core (k-core with $k = 3$) encircled.

In the following, we follow closely the derivation in [DGM06]. For an analysis of the Internet structure in terms of its different k-cores, see [CHK+07].

We now proceed to analyze the structure of the core. We start with a network having degree distribution $P(k)$ and assume that a fraction $q \equiv 1 - p$ of the nodes are randomly removed from the network. The network is locally tree-like. Therefore, we assume that if a node has at least C neighbors that have at least $C - 1$ similar infinite C-branches growing from them, then this node belongs to the giant C-core. Since the size of each of these branches grows exponentially, it is guaranteed that for every two nodes fulfilling this condition, the branches emanating from them will collide in all possible combinations (i.e., each of the C or more branches emanating from one node will collide with each of the C emanating from the other). Therefore, it is expected that any two such nodes will belong to the same component. In fact, since each branch is expected to collide with all other branches, it is expected that there will be C independent paths between the nodes.

Let us use R to denote the probability a given end of a link in the network is not the root of a $C - 1$ tree (that is, an infinite tree, which, in each layer, splits to at least $C - 1$ similar branches). A node belongs to the C-core if at least C of the links emanating from it are the roots of such a $C - 1$ tree. The probability of a node belonging to the C-core is:

$$M(C) = p \sum_{k \geq C} P(k) \sum_{n=C}^{k} \binom{k}{n} R^{k-n} (1 - R)^n \,. \tag{10.56}$$

The outer sigma sums over all possible node degrees of C and above. The inner sigma sums over the number of infinite branches, $C \leq n \leq k$ stemming from the node. See Figure 10.7 for a graphical presentation. Note that for $C = 1$, Eq. (10.56) reduces to ordinary percolation.

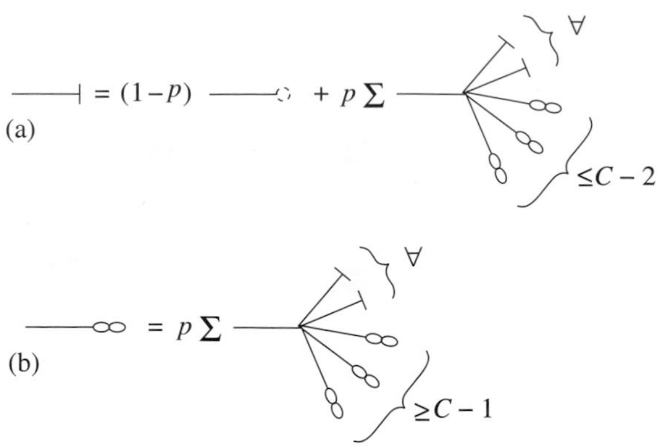

Figure 10.7 An illustration of (a) Eq. (10.56) and (b) Eq. (10.57). After [GDM06].

To calculate R, we note that for an end of a link not to be the root of an infinite $C-1$ tree, it must be deleted (with probability $1-p$) or have at most $C-2$ infinite children branches. The probability of each link reached leading to an infinite $C-1$ tree is $1-R$. Therefore, if a link leads to a node of degree $k = i+1$, then the probability that n of its i outgoing links will lead to infinite $C-1$ trees is $\binom{i}{n} R^{i-n}(1-R)^n$. This leads to the equation

$$R = 1 - p + p\sum_{n=0}^{C-2}\left[\sum_{i=n}^{\infty}\frac{(i+1)P(i+1)}{\langle k\rangle}\binom{i}{n}R^{i-n}(1-R)^n\right]. \tag{10.57}$$

Using the generating function from Chapter 9, we can rewrite Eq. (10.57) as

$$R = 1 - p + p\Phi_C(R), \tag{10.58}$$

with

$$\Phi_C(R) = \sum_{n=0}^{C-2}\frac{(1-R)^n}{n!}\frac{\mathrm{d}^n}{\mathrm{d}R^n}G_1(R)\,. \tag{10.59}$$

Equation (10.57) always has the trivial solution $R = 1$, indicating no giant C-core. The transition occurs at the point where another solution with $R < 1$ appears. At this point the solution $R = 1$ becomes unstable and a giant C-core forms.

Define the function

$$f_C(R) = \frac{1-\Phi_C(R)}{1-R}\,. \tag{10.60}$$

Using this function, Eq. (10.57) can be reformulated as

$$pf_C(R) = 1. \tag{10.61}$$

For a reasonably well-behaved degree distribution $P(k)$, $\Phi_C(R)$ is a continuous function, and therefore $f_C(R)$ is also continuous in the range $0 \leq R < 1$. Thus,

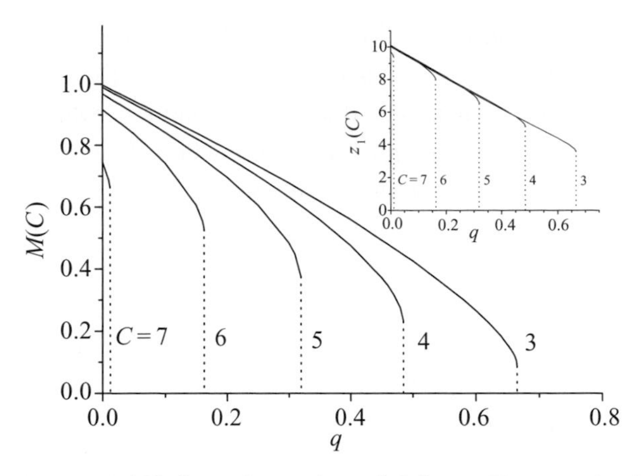

Figure 10.8 The size of the C-core, $M(C)$, for various values of C for an ER network with average degree $\langle k \rangle = 10$ as a function of the fraction of removed nodes, q. The ordinary percolation threshold in this case is $q_c = 0.9$. The inset shows the mean degree, $z_1(C)$, of nodes in the C-core. After [DGM06].

Eq. (10.61) has a non-trivial solution if

$$p \max_{R \in [0,1)} f_C(R) \geq 1 \,. \tag{10.62}$$

Exactly at the critical point the maximum of $f_C(R)$ touches the line $y = p_c^{-1}$. Therefore, we obtain two equations for the threshold:

$$p_c(C) = \frac{1}{f_C(R_{\max})}, \qquad f_C'(R_{\max}) = 0 \,. \tag{10.63}$$

These equations allow us to calculate both p_c, the critical threshold, and $R_{\max}$, the value R at the critical threshold. For ordinary percolation, which is equivalent to the cases $C = 1, 2$,[5] we are used to seeing the same value of the order parameter at both sides of the phase transition. That is, only the slope (derivative) of the order parameter changes at the transition point. The order parameter itself is continuous. This is typical of a second-order phase transition. This is not the case, however, in k-core percolation.

The k-core percolation is an example of a hybrid phase transition. It is a first-order phase transition, meaning that the order parameter (R or $M(C)$) changes discontinuously (see Figure 10.8). However, critical exponents can be identified near the transition. We will not discuss further properties of this transition. The interested reader is referred to [DGM06, GDM06] for further details. For completeness we will only discuss briefly the properties of k-core percolation in scale-free networks.

[5] $C = 1$ is identical to ordinary percolation, whereas $C = 2$ gives the size of the giant component without the dangling ends, i.e., without all branches leading to a dead end.

For scale-free networks with $\gamma > 3$ and minimum degree $m = 1$, no C-core exists for $C \geq 3$. As m increases, however, C-cores come into existence for various C where, naturally, if $m \geq C$ the entire network is its own C-core. For $2 < \gamma \leq 3$, a C-core exists for any m and any finite C, p. This stems from the divergence of $f_C(R)$ at $R \to 1$, meaning that a non-trivial solution of Eq. (10.61) exists for every p, C.

The analysis of the k-core in [CHK$^+$07] attempts to study the structure of the Internet through its C-cores for different values of C. An interesting twist there is the discussion of the network structure left *after removing the C-core for some C*. This is somewhat similar to the intentional attack strategy presented above. The nodes outside the C-core are very similar to the network after an attack on the highest degree nodes.

10.7 Conclusions

We reviewed results for the percolation properties of scale-free graphs, and have shown that some properties of such graphs differ from generalized random graphs owing to the diverging moment of the degree distribution. We have seen that scale-free graphs with $2 < \gamma < 3$ are very robust to random breakdown of almost 100% of the nodes, making this a favorable design for unmanaged networks, such as the Internet, where nodes can be disconnected or can fail unexpectedly. However, these networks are vulnerable to intentional attacks on the most highly connected nodes. The lack of percolation threshold also makes those networks sensitive to virus propagation [PV01a, PV01b].

We also described the behavior of networks near the percolation transition. It is shown that scale-free networks with large γ ($\gamma > 4$) near the percolation transition behave as expected in percolation in infinite-dimensional lattices. We have seen that near the threshold there is a crossover between a finite fractal dimension at short distances to infinite-dimensional, exponential growth at large distances. Thus, networks near the critical point become sparser, and communication becomes inefficient, as packets have to travel a long distance (many routers) on the way to their destination.

We have seen that the critical exponents behave differently in scale-free networks than in ordinary networks both in the regime $3 < \gamma < 4$, where a transition occurs at a finite p_c, and in the regime $2 < \gamma < 3$, where $p_c \to 0$. In particular, $\beta > 1$ in both regimes, making the transition to higher than second order. However, for $\gamma > 4$, the percolation exponents converge to the regular mean-field exponents, similarly to ER networks. This suggests that for certain properties scale-free networks can be viewed as a generalization of ER graphs.

Note that the attack on high-degree nodes is not the most efficient method for breaking a network. Several other methods have been devised that perform better than attack by degree and outperform random removal even for random regular networks, where all degrees are equal. Some of these methods are based on betweenness centrality or on adaptive estimation of a node's importance in the network during the attack. For a survey of such attack methods, see [Mag03]. Although the problem of determining the optimal attack on a network is believed to be computationally difficult, some approximations can be made, giving better results than attack by degree, and showing interesting phase transition properties [CPH+08, PCS+07]. The theory of percolation in correlated networks is studied in more detail in [GDM08].

Exercises

10.1 Find the percolation criterion when each node, of degree k is removed with probability $r(k)$. This type of percolation is discussed in [DB02, GCA+05] in conjunction with immunization and various types of attacks or failures.

10.2 Generalize the criterion obtained in Section 10.5 to the case where the probability of removing a node depends on k, $p = p(k)$.

10.3 A network has nodes of degrees 2 and 3, where $P(k_1 = 2|k_2 = 3) = 1$, i.e., nodes are only connected to nodes of the other degree.
(a) Deduce the degree distribution $P(k)$.
(b) Use the matrix formulation to deduce the percolation threshold of this network. Can the threshold be obtained using a more direct method?

10.4 Using the result of Exercise 4.6 calculate κ for a random geometric graph. Use the result to estimate the percolation threshold for a random geometric graph. Is this an over-estimate or an under-estimate? Explain.

10.5 Generalize the criterion obtained in Section 10.5 to the case of bond percolation, where the probability of removing a link (bond) depends on both degrees, k and k', $p = p(k, k')$.

10.6 Obtain the full expression for the product $\vec{v}^t A^2 \vec{v}$ in Section 10.5 and show that it diverges.

10.7 Find the percolation threshold for a bipartite network (see Exercise 4.2 for the definition) with degree distributions $P_a(k)$ and $P_b(k)$. Consider the following cases.

(a) Only nodes of type a are removed.

(b) Nodes of type a are removed with probability p_a and nodes of type b are removed with probability p_b.

10.8 Obtain the equations for the critical threshold of k-core percolation in ER networks. That is, find the minimum $\langle k \rangle$ for which a K-core exists for some K. The numerical results for the solution appear in Figure 10.8.

11 Structure of random directed networks: the bow tie

Many real complex networks have directed links, a property that affects the network's navigability and large-scale topology. Here we present the percolation properties of such directed scale-free networks with correlated *in*- and *out*-degree distributions [SCbA$^+$02]. We derive a phase diagram that indicates the existence of three regimes, determined by the values of the degree exponents. In the first regime, we attain the known directed percolation mean-field exponents. In contrast, the second and third regimes are characterized by anomalous mean-field exponents that we calculate analytically. In the third regime the network is resilient to random dilution, i.e., the percolation threshold is $p_c \to 0$.

11.1 Introduction

Real networks are sometimes directed; for example, in social and economical networks, a node A gains information or acquires physical goods from node B, but node B does not necessarily get similar input from node A. Likewise, most metabolic reactions are one directional. Thus, changes in the concentration of molecule A affect the concentration of its product B, but the reverse is not true. Despite the directedness of many real networks, the modeling literature, with few notable exceptions [CRS$^+$03, DMS01b, NSW01, SCbA$^+$02], has focused mainly on undirected networks.

An important property of directed networks can be illustrated by studying their degree distribution, $P(j, k)$, or the probability that an arbitrary node has j incoming and k outgoing edges. Many natural directed networks, such as the WWW, metabolic networks, citation networks, and others, exhibit a power-law, or *scale-free* degree distribution for the incoming and/or outgoing links:

$$P_{\text{in(out)}}(k) = ck^{-\gamma_{\text{in(out)}}}, \quad k \geq m , \tag{11.1}$$

where m is the minimal connectivity (usually taken to be $m = 1$), c is a normalization factor and $\gamma_{\text{in(out)}}$ are the in(out)-degree exponents characterizing the network

[BAJ00, BKM$^+$00]. An important property of scale-free networks is their robustness to random failures, coupled with an increased vulnerability to attacks [AJB00, CEbH00, CEbH01, CNSW00, SM01]. It has been recognized that this feature can be addressed analytically in quantitative terms [CEbH00, CEbH01, CNSW00] by combining graph theoretical concepts with ideas from percolation theory (see Chapter 10). Yet, although the percolation properties of undirected networks have been extensively studied, little is known about the effect of node failure in directed networks. Since many important networks are directed, it is important to understand the implications regarding their stability. Here we show that directedness has a strong impact on the percolation properties of complex networks and we will draw a detailed phase diagram.

11.2 Structure

The structure of a directed graph has been characterized in [DMS01b, NSW01], and in the context of the WWW in [BKM$^+$00]. In general, a directed graph consists of a giant weakly connected component (GWCC) and several finite components. In the GWCC every node is reachable from every other, provided that the links are treated as bidirectional. The GWCC is further divided into a giant strongly connected component (GSCC), consisting of all nodes reachable from each other following directed links. All the nodes reachable from the GSCC are referred to as the giant OUT component, and the nodes from which the GSCC is reachable are referred to as the giant IN component. The GSCC represents the intersection of the IN and OUT components. All nodes in the GWCC, but not in the IN and OUT components are referred to as the "tendrils" (see Figure 11.1).

11.3 The giant component

For a directed random network of arbitrary degree distribution, the condition for the existence of a giant component can be deduced in a manner similar to that shown in Chapter 9. Assume that the network has a degree distribution $P(j, k)$, representing the joint probability of having in-degree j and out-degree k. If a node is reached following a link pointing to it, then it must have at least one outgoing link, on average, in order to be part of a giant component. Assuming a link point from node a to node b, this

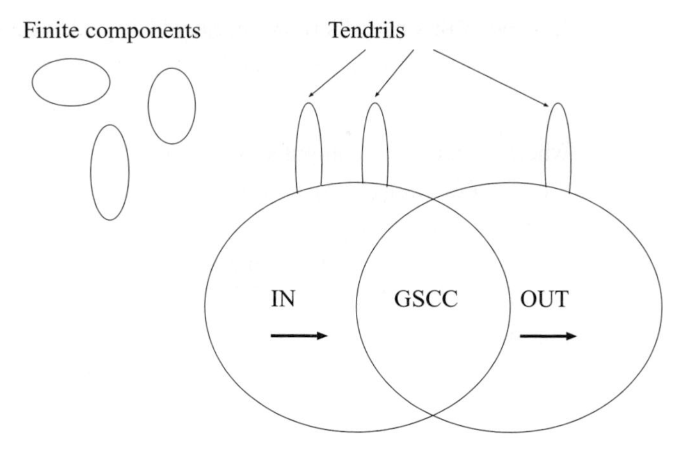

Figure 11.1 Structure of a general directed graph.

condition can be written as

$$\langle j_b | a \to b \rangle = \sum_{k_b, j_b} j_b P(j_b, k_b | a \leftrightarrow b) = 1. \tag{11.2}$$

Using Bayes' rule (see Appendix A), we obtain

$$\begin{aligned} P(j_b, k_b | a \to b) &= P(j_b, k_b, a \leftrightarrow b) / P(a \to b) \\ &= P(a \to b | j_b, k_b) P(j_b, k_b) / P(a \to b) \end{aligned}$$

for random networks $P(a \to b) = \langle k \rangle / (N-1)$ and $P(a \to b | j_b, k_b) = k_a/(N-1)$, where N is the total number of nodes in the network. The above criterion thus reduces to [DMS01b, NSW01]

$$\langle jk \rangle \geq \langle k \rangle . \tag{11.3}$$

11.4 Percolation in directed scale-free networks

Suppose a fraction q of the nodes is removed from the network. (Alternatively, a fraction $p = 1 - q$ of the nodes is retained.) The original degree distribution, $P(j, k)$, becomes

$$\begin{aligned} P'(j, k) = \sum_{j_0, k_0}^{\infty} P(j_0, k_0) \binom{j_0}{j} p^j (1-p)^{j_0 - j} \\ \times \binom{k_0}{k} p^k (1-p)^{k_0 - k} . \end{aligned} \tag{11.4}$$

In view of this new distribution, Eq. (11.3) yields the percolation threshold

$$p_c = 1 - q_c = \frac{\langle k \rangle}{\langle jk \rangle}, \tag{11.5}$$

where averages are computed with respect to the original distribution before dilution, $P(j,k)$. Equation (11.5) indicates that in directed scale-free networks if $\langle jk \rangle$ diverges, then $p_c \to 0$ and the network is resilient to random breakdown of nodes and links.

The term $\langle jk \rangle$ may be dramatically influenced by the appearance of correlations between the in- and out-degrees of the nodes. In particular, let us consider scale-free distributions for both the in- and out-degrees

$$P_{\text{in}}(j) \sim \begin{cases} Bc_{\text{in}} j^{-\gamma_{\text{in}}} & j \neq 0 \\ 1 - B & j = 0, \end{cases} \tag{11.6}$$

and

$$P_{\text{out}}(k) = c_{\text{out}} k^{-\gamma_{\text{out}}}. \tag{11.7}$$

In (11.6) we choose to add the possibility of zero value to the in-degree in order to maintain $\langle j \rangle = \langle k \rangle$. If the in- and out-degrees are uncorrelated, we expect $\langle jk \rangle = \langle j \rangle \langle k \rangle$. For several real directed networks, this equality does not hold. For example, the network of the Notre Dame University WWW [BAJ00] has $\langle k \rangle = \langle j \rangle \approx 4.6$, and thus $\langle j \rangle \langle k \rangle = 21.16$. In contrast, measuring directly yields $\langle jk \rangle \approx 200$, about an order of magnitude larger than the result expected for the uncorrelated case. This yields, using Eq. (11.5), an estimate of $q_c \approx 0.02$, i.e., a very stable directed network. Similar results have also been obtained for some metabolic networks [JTA$^+$00], indicating that in real directed networks, the in- and out-degrees are correlated [SCbA$^+$02].

To address correlations, we model it as follows. We first generate the j values for the entire network. Next, for each node with $j \neq 0$ with probability A, we generate k fully correlated with j, i.e., $k = k(j)$. Assuming that $k(j)$ is a monotonically increasing function, then the requirement $c_{\text{out}} k^{-\gamma_{\text{out}}} \mathrm{d}k = c_{\text{in}} j^{-\gamma_{\text{in}}} \mathrm{d}j$, which is needed to maintain scale-free distributions, leads to $k^{\gamma_{\text{out}}-1} = j^{\gamma_{\text{in}}-1}$. With probability $1 - A$, the degree k is chosen independently of j:

$$P(j,k) \sim \begin{cases} (1-A)Bc_{\text{in}} j^{-\gamma_{\text{in}}} c_{\text{out}} k^{-\gamma_{\text{out}}} & \\ \qquad + BAc_{\text{out}} k^{-\gamma_{\text{out}}} \delta_{j,j(k)} & j \neq 0 \\ (1-B)c_{\text{out}} k^{-\gamma_{\text{out}}} & j = 0, \end{cases} \tag{11.8}$$

where $j(k) = k^{\gamma_{\text{out}}-1/\gamma_{\text{in}}-1}$. With this distribution, any finite fraction BA of fully correlated nodes yields a diverging $\langle jk \rangle$ whenever

$$(\gamma_{\text{out}} - 2)(\gamma_{\text{in}} - 2) \leq 1, \tag{11.9}$$

causing the percolation threshold to vanish (see Figure 11.2).

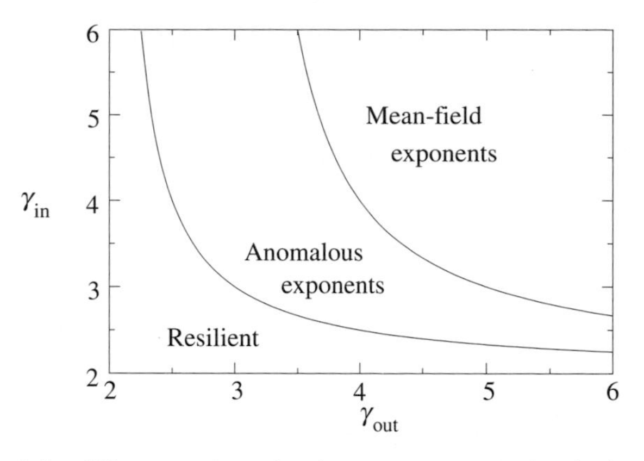

Figure 11.2 Phase diagram of the different regimes for the IN component of scale-free correlated directed networks. The boundary between resilient and anomalous exponents is derived from Eq. (11.9), whereas that between anomalous exponents and mean-field exponents is given by Eq. (11.24) for $\gamma^\star = 4$. For the diagram of the OUT component, γ_{in} and γ_{out} change roles. Taken from [SCbA+02].

In the case of no correlations between the in- and the out-degrees, $A = 0$, Eq. (11.8) becomes $P(j, k) = P_{in}(j)P_{out}(k)$. Then the condition for the existence of a giant component is: $\langle k\rangle = \langle j\rangle = 1$. Moreover, Eq. (11.5) reduces to:

$$p_c = 1 - q_c = \frac{1}{\langle k\rangle} . \tag{11.10}$$

Applying (11.10) to scale-free networks, one can conclude that for $\gamma_{out} > 2$ and $\gamma_{in} > 2$ a phase transition exists at a finite q_c. Here we concern ourselves with the critical exponents associated with the percolation transition in a scale-free network with $\gamma_{out} > 2$ and $\gamma_{in} > 2$, which is the most relevant regime (Figure 11.2).

Thus, percolation of the GWCC can be seen as similar to percolation in the non-directed graph created from the directed graph by ignoring the directionality of the links. The threshold is obtained from the criterion [CEbH00]

$$p_c = \frac{\langle k\rangle}{\langle k(k-1)\rangle}. \tag{11.11}$$

Here the degree distribution is the convolution of the in and out distributions

$$P'(k) = \sum_{l=0}^{k} P(l, k-l). \tag{11.12}$$

Regardless of correlations, $P'(k)$ is always dominated by the slower decay-exponent; therefore, percolation of the GWCC is the same as in non-directed scale-free networks, with $\gamma_{eff} = \min(\gamma_{in}, \gamma_{out})$. Note that the percolation threshold of the GWCC may differ from that of the GSCC and the IN and OUT components [DMS01b].

11.5 Critical exponents

We now use the formalism of generating functions [Wil94] to analyze percolation of the GSCC and IN and OUT components. In [DMS01b, NSW01] a generating function is built for the joint probability distribution of outgoing and incoming degrees, before dilution:

$$\Phi(x, y) = \sum_{k,j} P(j, k) x^j y^k \,. \tag{11.13}$$

Using the approach of Callaway *et al.* [CNSW00], let $p(j, k)$ be the probability that a vertex of degree (j, k) remains in the network following dilution. The generating function after dilution is then

$$G(x, y) = \sum_{k,j} P(j, k) p(j, k) x^j y^k \,. \tag{11.14}$$

From (11.14), it is possible to define the generating function for the outgoing degrees G_0

$$G_0(y) \equiv G(1, y) = \sum_{k,j} P(j, k) q(j, k) y^k \,. \tag{11.15}$$

The probability of reaching a node by following a specific link is proportional to $jP(j, k)$. Therefore, the probability of reaching an occupied node following a specific directed link is generated by

$$G_1(y) = \frac{\sum_{j,k} jP(j, k) q(j, k) y^k}{\sum_{j,k} jP(j, k)} \,. \tag{11.16}$$

Let $H_1(y)$ be the generating function for the probability of reaching an outgoing component of a given size by following a directed link, after a dilution. $H_1(y)$ satisfies the self-consistent equation:

$$H_1(y) = 1 - G_1(1) + yG_1(H_1(y)) \,. \tag{11.17}$$

Since $G_0(y)$ is the generating function for the outgoing degree of a node, the generating function for the probability that n nodes are reachable from a given node is

$$H_0(y) = 1 - G_0(1) + yG_0(H_1(y)) \,. \tag{11.18}$$

For the case where correlations exist, and assuming random dilution, $q(j, k) = q$, Eqs. (11.17) and (11.18) reduce to

$$H_1(y) = 1 - q + \frac{qy}{\langle j \rangle} \sum_k (BAj(k) + (1 - A)\langle j \rangle) P_{\text{out}}(k) H_1(y)^k \,, \tag{11.19}$$

and

$$H_0(y) = 1 - p + py \sum_k P_{\text{out}}(k) H_1(y)^k . \tag{11.20}$$

If $A \to 0$, one expects that $H_0(y) = H_1(y)$, since there is no correlation between j and k. Thus, the probability of having k outgoing links is $P_{\text{out}}(k)$, whether we choose the node randomly or weighted by the incoming edges j.

Here, $H_0(1)$ is the probability of reaching an outgoing component of any *finite* size choosing a node. Thus, below the percolation transition $H_0(1) = 1$, whereas above the transition there is a finite probability of following a directed link to a node that is a root of an infinite outgoing component: $P_\infty = 1 - H_0(1)$. It follows that

$$P_\infty(p) = p\left(1 - \sum_k^\infty P_{\text{out}}(k) u^k\right) , \tag{11.21}$$

where $u \equiv H_1(1)$ is the smallest positive root of

$$u = 1 - p + \frac{p}{\langle j \rangle} \sum_k (BAj(k) + (1 - A)\langle j \rangle) P_{\text{out}}(k) u^k . \tag{11.22}$$

Here, $P_\infty(p)$ is the fraction of nodes from which an infinite number of nodes is reachable. Equation (11.22) can be solved numerically and the solution can be substituted into Eq. (11.21), yielding the size of the OUT component at dilution $q = 1 - p$.

11.5.1 Size of the giant component

Near criticality, the probability of starting from a node and reaching the giant outgoing component follows $P_\infty \sim (p - p_c)^\beta$. For mean-field systems, such as infinite-dimensional systems, random graphs, and Cayley trees, it is known that $\beta = 1$ (see Chapter 10) [BH96, FHL01, SA94]. However, this regular mean-field result is not always valid for scale-free networks. Instead, for certain values of γ_{in} and γ_{out}, anomalous mean-field exponents appear owing to the heterogeneity in the node degrees.

Following [CbH02], we study the singular behavior of Eq. (11.22) near $p = p_c$, $u = 1$, and find

$$\beta = \begin{cases} \frac{1}{3-\gamma^\star} & 2 < \gamma^\star < 3 \\ \frac{1}{\gamma^\star - 3} & 3 < \gamma^\star < 4 \\ 1 & \gamma^\star > 4, \end{cases} \tag{11.23}$$

where

$$\gamma^\star = \gamma_{\text{out}} + \frac{\gamma_{\text{in}} - \gamma_{\text{out}}}{\gamma_{\text{in}} - 1} . \tag{11.24}$$

We see that the order parameter exponent β attains its usual mean-field value only for $\gamma^\star > 4$. Since $\gamma_{\text{out}} \to \gamma_{\text{in}}$, the correlated fraction BA of nodes resembles non-directed networks [CbH02, PV01a] (where there is no distinction between incoming and outgoing degrees). In this case, we obtain $\gamma^\star = \gamma_{\text{out}} = \gamma_{\text{in}}$ for any amount of correlation A. The criterion for the existence of a giant component is then $\langle k^2 \rangle / \langle k \rangle = 1$, and not 2 as in the non-directed case. The difference stems from the fact that in the non-directed case, one of the links is used to reach the node, whereas in the directed case there is generally no correlation between the location of the incoming and outgoing links. Therefore, one more outgoing link is available for leaving the node.

Without any correlations, $A = 0$, different terms prevail in the analysis and

$$\beta = \begin{cases} \frac{1}{\gamma_{\text{out}}-2} & 2 < \gamma_{\text{out}} < 3 \\ 1 & \gamma_{\text{out}} > 3. \end{cases} \tag{11.25}$$

This is the same as Eq. (11.23), but with $\gamma^\star = \gamma_{\text{out}} + 1$.

The GSCC is the intersection of the IN and OUT components. Therefore, it behaves as the smaller of the two components: $\beta_{\text{GSCC}} = \max(\beta_{\text{in}}, \beta_{\text{out}})$. This can also be derived by applying the same methods as for the IN and OUT components to the generating function of the GSCC obtained in [DMS01b]. The exponent for the GWCC, on the other hand, is independent of the exponents of the other components, since the transition point is different.

11.5.2 Finite component sizes

It is known that for a random graph of arbitrary degree distribution the finite clusters follow the scaling form [BH96, SA94]

$$n(s) \sim s^{-\tau} e^{-s/s^*}, \tag{11.26}$$

where s is the cluster size and $n(s)$ is the number of clusters of size s. At criticality $s^* \sim |p - p_c|^{-\sigma}$ diverges and the tail of the distribution follows a power law.

The probability that s nodes can be reached from a node by following links at criticality follows $p(s) \sim s^{-\tau}$, and is generated by H_0, where $H_0(y) = \sum_s p(s) y^s$. As in [CbH02], $H_0(y)$ can be expanded from Eq. (11.18). In the presence of correlations, we find

$$\tau = \begin{cases} 1 + \frac{1}{\gamma^\star-2} & 2 < \gamma^\star < 4 \\ \frac{3}{2} & \gamma^\star > 4. \end{cases} \tag{11.27}$$

Table 11.1 Values of $\gamma^\star$ for the different network components for both correlated and uncorrelated cases, after [SCbA+02]

	Uncorrelated	Correlated
GWCC	$\min(\gamma_{out}, \gamma_{in}) + 1$	$\min(\gamma_{out}, \gamma_{in})$
in	$\gamma_{out} + 1$	$\gamma_{out} + \frac{\gamma_{in} - \gamma_{out}}{\gamma_{in} - 1}$
out	$\gamma_{in} + 1$	$\gamma_{in} + \frac{\gamma_{out} - \gamma_{in}}{\gamma_{out} - 1}$
GSCC	$\min(\gamma_{out}, \gamma_{in}) + 1$	$\min(\gamma_{out}^*, \gamma_{in}^*)$

The regular mean-field exponents are recovered for $\gamma^\star > 4$. For the uncorrelated case, we get

$$\tau = \begin{cases} 1 + \frac{1}{\gamma_{out} - 1} & 2 < \gamma_{out} < 3 \\ \frac{3}{2} & \gamma_{out} > 3. \end{cases} \tag{11.28}$$

Now the regular mean-field results are obtained for $\gamma > 3$.

11.6 Summary

In summary, we calculated the percolation properties of directed scale-free networks. We found that the percolation threshold and the critical exponents in scale-free networks depend strongly upon the existence of correlations and upon the degree distribution exponents in the range of $2 < \gamma^\star < 4$. This regime characterizes most naturally occurring directed networks, such as metabolic networks or the WWW. The regular mean-field behavior of percolation in infinite dimensions is recovered only for $\gamma^\star > 4$. A connection is found between non-directed and directed scale-free percolation exponents for any finite correlation between the in- and out-degrees. In the uncorrelated case, i.e., $P(j, k) = P_{in}(j)P_{out}(k)$, the probability of reaching an outgoing component does not have any dependence upon $P_{in}(j)$. The results are summarized in Table 11.1.

Exercises

11.1 For a network with all nodes having degrees $j = k = c$ for some constant c, what is the percolation threshold $p_c(c)$?

11.2 Given a random directed network with given distributions $P_{\mathrm{in}}(j)$ and $P_{\mathrm{out}}(k)$ and no correlations, consider removal of nodes with the probability depending on both the in- and out-degrees, $P(j, k)$. What is the condition for the existence of a giant component in this case?

11.3 (Research) Study intentional attacks on random directed networks. Find which attacks (based on some combination of the in- and out-degrees) lead to the fastest breakdown of the network.

11.4 (Research) Study the upper critical dimension in percolation on random directed networks.

11.5 (Research) Study the chemical and fractal dimensions in percolation on random directed networks.

12 Introducing weights: bandwidth allocation and multimedia broadcasting

12.1 Introduction

In this chapter we present studies of the optimal distance in networks, ℓ_{opt}, defined as the length of the path minimizing the total weight, in the presence of disorder [BBC$^+$03]. Disorder is introduced by assigning random weights to the links or nodes. These weights may represent properties of the links and the transport in them. These properties may include the bandwidth of links in communication networks, delays in the transport of data or material, or the cost of traversing the link. Many real-world networks may be better described by introducing link weights, an example is the airline network where the frequency of flights is an important property (see, e.g., [BBPV04]).

An important quantity characterizing networks is the average distance (minimal hopping) ℓ_{min} between two nodes in a network containing N nodes. For the Erdős–Rényi networks [ER59, ER60], and the related, more realistic Watts–Strogatz (WS) network [WS98], ℓ_{min} scales as $\ln N$ [Bol85], which leads to the concepts of small world and "six degrees of separation," while for scale-free networks ℓ_{min} scales as $\ln \ln N$. See Chapter 6 for details.

In most studies, all links in the network are regarded as identical and thus the relevant parameter for information flow including efficient routing, searching, and transport is ℓ_{min}. In practice, however, the weights (for example, the quality or cost) of links are usually not equal, and thus the length of the optimal path, ℓ_{opt}, minimizing the sum of weights is usually longer than the distance ℓ_{min}. In many cases, the selection of the path is controlled by the sum of weights (for example, total cost) and this case corresponds to regular or weak disorder. However, in other cases, for example, when transmission at a constant high rate is needed (for example, in broadcasting video records over the Internet), the narrowest band link in the path between the transmitter and receiver controls the rate of transmission. This situation – in which one link controls the selection of the path – is called the strong disorder limit. Another example of strong disorder is when the distribution of weights is so broad that a single weight dominates the sum. In this chapter we show that disorder or inhomogeneity in the weight of links may increase the distance ℓ_{opt} dramatically, destroying the "small-world" nature of the networks. A similar effect is observed in the minimum spanning tree (or MST), i.e., the set of links with the minimum total weight needed to keep all nodes connected. In the minimum spanning tree, the optimization is global, and the lowest weight tree for the entire network is chosen, leading to increased distances, like in strong disorder, but for any weight distribution, not necessarily a broad one.

12.2 Random weighted networks

To implement the disorder, we assign a weight or "cost" to each link or node in the network. For example, the weight could be the time τ_i required to transit the link i. The optimal path connecting nodes A and B is the one for which $\sum_i \tau_i$ is a minimum. Whereas in weak disorder, all links contribute to the sum, in strong disorder, one term dominates it. The strong disorder limit may be realized naturally in the vicinity of the absolute zero temperature if passing through a link is an activation process with a random activation energy ϵ_i and $\tau_i = \exp(\beta\epsilon_i)$, where β is the inverse temperature [CLHS06]. Let us assume that the energy spectrum is discrete and that the minimal difference between energy levels is $\Delta\epsilon$. It can be easily shown that if $\beta > \ln 2/\Delta\epsilon$, the value of $\sum_i \tau_i$ is dominated by the largest term, $\tau_{\max}$. Thus, if we have two different paths characterized by the sums $\sum_i \tau_i$ and $\sum_i \tau_i'$, such that $\tau_{\max} > \tau_{\max}'$, it follows that $\sum_i \tau_i > \sum_i \tau_i'$.

To generate ER graphs, we start with zN links and for each link randomly select from the total $N(N-1)/2$ possible pairs of nodes, a pair that is connected by this link. The WS network [WS98] is implemented by placing the N nodes on a circle. Initially, each node i is connected with z nodes $i+1, i+2, \ldots, i+z$ and periodic boundaries are implemented. Thus, each node has a degree $2z$ and the total number of links is zN. Next, we randomly remove a fraction p of the links and use them to connect randomly selected pairs of nodes. When $p = 1$, we obtain a model very similar to the ER graph.

To generate scale-free (SF) graphs, we employ the configuration model algorithm (see Appendix C) in which each node is first assigned a random integer k from a power-law distribution $P(k) \sim k^{-\gamma}$. Next, we randomly select a node and try to connect each of its k links with randomly selected k nodes that still have free positions for links.

We expect that the optimal path length in the weak disorder case will not be considerably larger than the shortest path, as found for regular lattices [KRW88] and random graphs [vdHHvM01]. Thus, we expect that the scaling for the shortest path $\ell_{\min} \sim \ln N$ will also be valid for the optimal path in weak disorder, but with a different prefactor, depending on the details of the graph and the distribution of weights. This, however, seems to change in the case of $2 < \gamma < 3$, where the ultra small-world effect may vanish when even a weak disorder is introduced.

In the case of strong disorder, we present the following theoretical arguments. Cieplak *et al.* [CMB99] showed that finding the optimal path between nodes A and B in the strong disorder limit is equivalent to the following procedure. First, we sort all M links of the network in descending order of their weights, so that the first link in this list has the largest weight. Since the sum of the weights on any path between nodes A and B is dominated by a single link with the largest weight, the optimal path cannot go through the first link in the list, provided there is a path between A and B that avoids this link. Thus, the first link in the list can be eliminated and now our problem is reduced to the problem of finding the minimal path on the network of $M - 1$ links. We can continue to remove links from the top of the list one-by-one until we pick a link whose removal destroys the connectivity between A and B. This means that all the remaining paths between A and B go through this singly connecting or "red" link [Con89] and that all these paths have the same largest weight corresponding to the "red" link. To continue optimization among these paths, we must select those paths with the minimal second largest term, the minimal third largest term and so on. Thus, we must continue to remove links in descending order of their weights unless they are "red" until a single path between A and B, consisting of only "red" links, remains. Since the assigning of weights to the links is random, so is their ordering. Hence, the optimization procedure in the strong disorder limit is statistically equivalent to removing the links in random order (without caring at all about weights) unless the connectivity between nodes A and B is destroyed. If connectivity is lost by removing some link, the link is restored and we continue with the next link until a single path exists. An illustration of the process can be seen in Figure 12.1.

At the beginning of this process, the chances of losing connectivity by removing a random link are very low, so the process corresponds exactly to diluting the network, which is identical to the percolation model. Only when the concentration of the remaining links approaches the percolation threshold will the chances of removing a singly connected "red" link [Con89] become significant, indicating that the optimal path must be on the percolation backbone connecting A and B. Since the network is not embedded in space but has infinite dimensionality, we expect from percolation theory that loops are not relevant at criticality [CEbH00]. Thus, the shortest path on the backbone must also be the optimal path.

We begin by considering the case of the ER graph that, at criticality, is equivalent to percolation on a Cayley tree or percolation at the upper critical dimension $d_c = 6$. For the ER graph, it is known that the mass of the incipient infinite cluster S scales as $N^{2/3}$ [ER59]. This result can also be obtained in the framework of percolation theory for $d_c = 6$. Since $S \sim R^{d_f}$ and $N \sim R^d$ (where d_f is the fractal dimension and R the

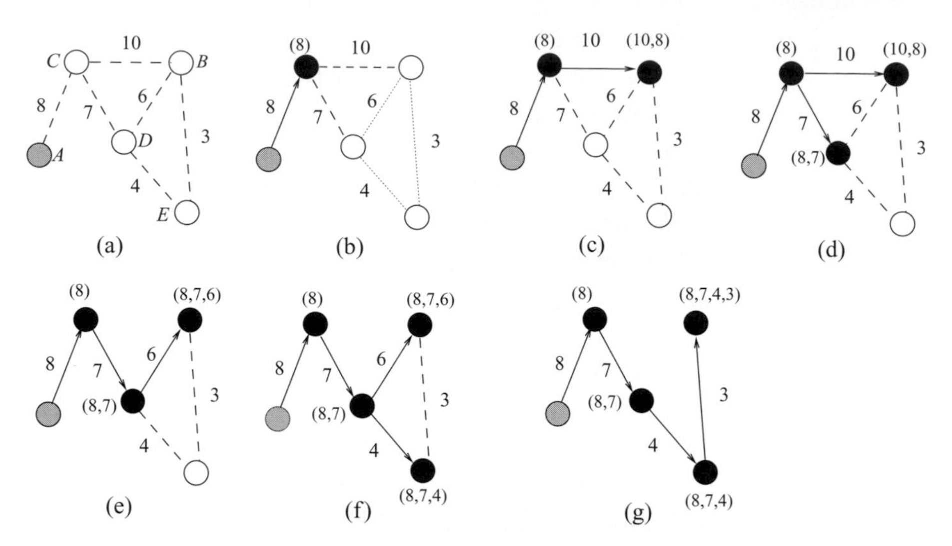

Figure 12.1 In (a) we show schematically a network consisting of five nodes (A, B, C, D, and E). The links between them are shown by dashed lines. The origin (A) is denoted in gray. All links were assigned random weights, shown beside the links. In (b) one node (C) has been visited for the first time (marked in black) and assigned the sequence (8) of length $\ell = 1$. The path is denoted by a solid arrow. Note that there is no other path going from the origin (A) to this node (C) so $\ell_{\text{opt}} = 1$ for that path. In (c) another node (B) is visited for the first time (marked in black) and is assigned the sequence $(10, 8)$ of length 2. The sequence has the information of all the weights of that path arranged in decreasing order. In (d) another node (D) is visited for the first time and is assigned the sequence $(8, 7)$ of length 2. In (e) node B, visited in (c) with sequence $(10, 8)$, is visited again with sequence $(8, 7, 6)$. The last sequence is smaller than the previous sequence $(10, 8)$ so node B is reassigned the sequence $(8, 7, 6)$ of length 3. The new path is shown as a solid line. In (f) a new node (E) is assigned with sequence $(8, 7, 4)$. In (g) node B is reached for the third time and is reassigned the sequence $(8, 7, 4, 3)$ of length 4. The optimal path for this configuration from A to B is denoted by the solid arrows in (g). After [HBB+05].

diameter of the cluster), it follows that $S \sim N^{d_{\mathrm{f}}/d}$ and thus for $d_{\mathrm{c}} = 6$, $d_{\mathrm{f}} = 4$ [BH96]

$$S \sim N^{2/3}. \tag{12.1}$$

It is also known [BH96] that, at criticality, at the upper critical dimension, $S \sim \ell^{d_\ell}$ where ℓ is the length of the shortest path on the cluster (component) and $d_\ell = 2$. Thus,

$$\ell \sim \ell_{\text{opt}} \sim S^{1/d_\ell} \sim N^{2/3d_\ell} \sim N^{\nu_{\text{opt}}}, \tag{12.2}$$

where $\nu_{\text{opt}} = 2/3d_\ell = 1/3$. We expect that the WS model [WS98] for large N and large p will be in the same universality class as ER.

The scaling relation in Eq. (12.2) is rigorous only as a lower bound. In order to establish that the optimal distance actually scales as $N^{1/3}$, it should be shown that

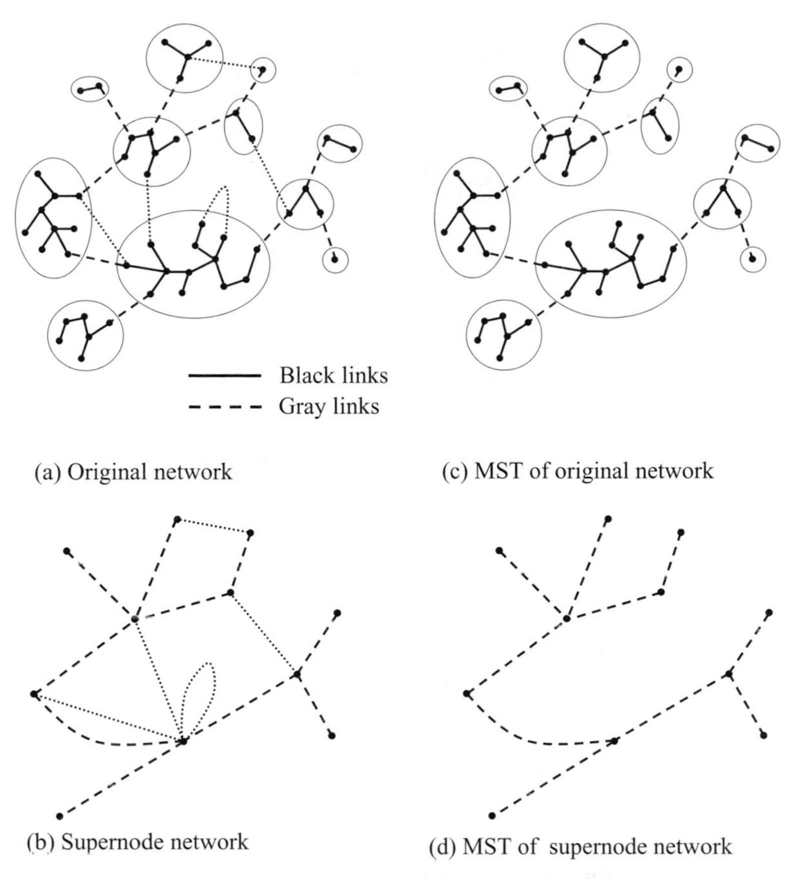

Figure 12.2 The structure of the minimum spanning tree. Below the percolation threshold the network consists of a small component with (almost) no loops. The links whose weights are below the threshold are called *black links*. These components consist of all parts of the minimum spanning tree. To connect the components, we must add links above the percolation threshold. These links are called the *gray links*, and only those that form no loops are added. After [BWC+07].

the percolation components connect in a compact way, i.e., that a small number of components is traversed in the optimal path between nodes (see Figure 12.2). A study of the behavior of these paths is presented in [KBB+05, WBHS06]. In [ABR06] a tight bound is finally established on the optimal path length for MSTs on the complete graph and ER networks. To establish the compactness of the connections between components of the network, one divides the links into three regimes. Up to the critical concentration, all links that do not form loops are added, leading to a critical network with $\mathcal{O}(N^{1/3})$ path length as presented above, Eq. (12.2). Then, a series of a few steps takes place in which the size of the largest component grows from $\mathcal{O}(N^{2/3})$ to $\mathcal{O}(N/\ln N)$. In every such step the length of the optimal path does

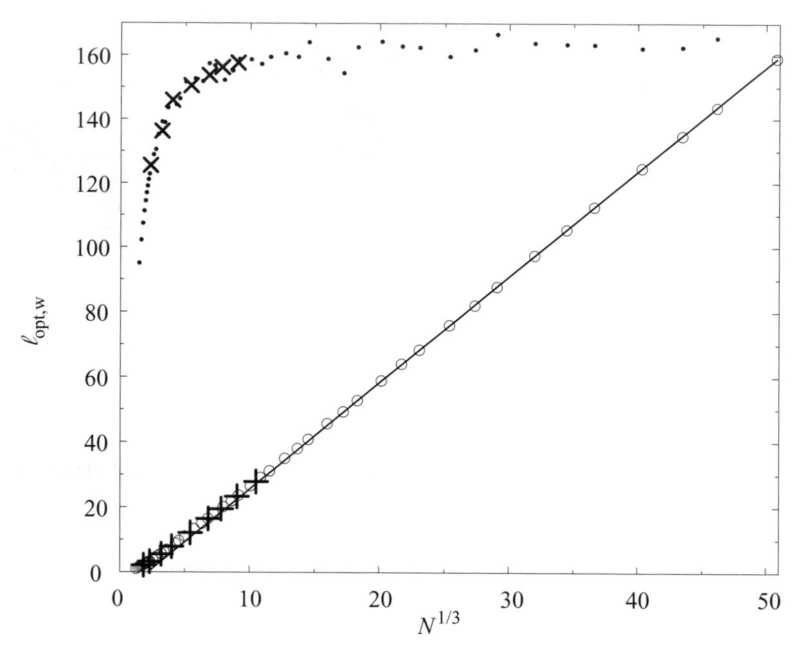

Figure 12.3 The dependence of ℓ_{opt} on $N^{1/3}$ for ER graphs for the strong disorder case obtained by direct optimization (+) and by randomly removing links (○). The linear asymptote has a slope of 3.27. Also shown are the successive slopes multiplied by 50 for direct optimization (×) and for randomly removing links (•). After [BWC+07].

not increase too much. Eventually, after reaching a component of size $\mathcal{O}(N/\ln N)$, all other components are considerably smaller and then connect to the largest component through a short sequence of small components. For full details, see [ABR06].

For scale-free networks, we can also use the percolation results at criticality. As discussed in Chapter 10, $d_\ell = 2$ for $\gamma > 4$, $d_\ell = (\gamma - 2)/(\gamma - 3)$ for $3 < \gamma < 4$, $S \sim N^{2/3}$ for $\gamma > 4$, and $S \sim N^{(\gamma-2)/(\gamma-1)}$ for $3 < \gamma \le 4$. Hence, we can conclude that

$$\ell \sim \ell_{opt} \sim \begin{cases} N^{1/3} & \gamma > 4 \\ N^{(\gamma-3)/(\gamma-1)} & 3 < \gamma \le 4. \end{cases} \tag{12.3}$$

In Figure 12.3 we present numerical simulations in the strong disorder limit by randomly removing links (or nodes) for ER networks. We also present additional simulations for the case of strong disorder on ER networks using direct optimization [CLHS06]. As can be seen, the results are identical to the results obtained by randomly removing links.

For scale-free networks, the behavior of the optimal path in the weak disorder limit is shown in Figure 12.4 for different degree distribution exponents γ. Here we plot ℓ_{opt} as a function of $\ln N$. All the curves have linear asymptotes, but the slopes depend

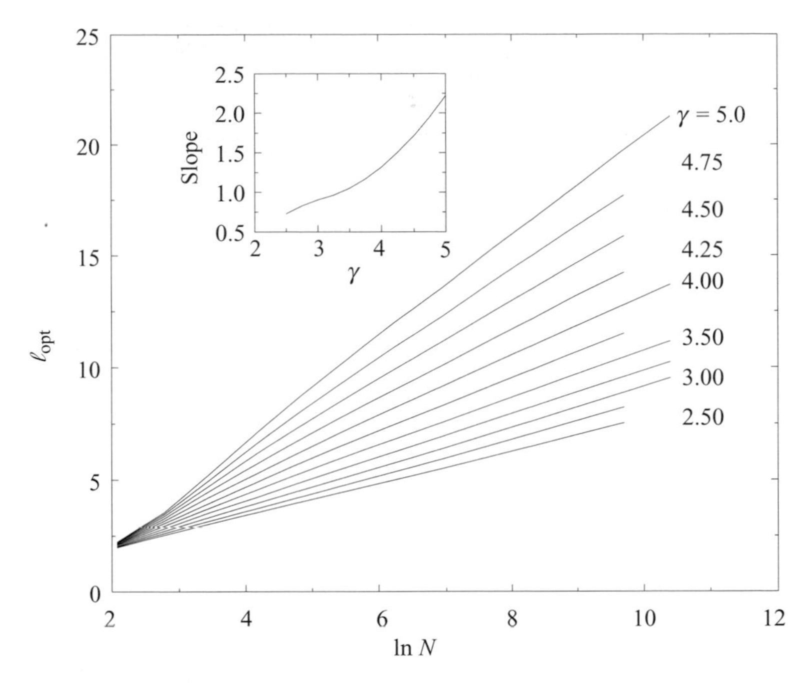

Figure 12.4 The dependence of ℓ_{opt} on $\ln N$ for scale-free graphs in the weak disorder case for various values of γ is shown on the graph. The behavior of the asymptotic slope versus γ is shown as an inset. After [BWC+07].

on γ,

$$\ell_{\text{opt}} \sim f(\gamma) \ln N. \tag{12.4}$$

This result is analogous to the behavior of the shortest path $\ell \sim \ln N$ for $\gamma > 3$. However, for $2 < \gamma < 3$, ℓ scales as $\ln \ln N$ [CH03], whereas ℓ_{opt} is significantly larger and scales as $\ln N$. This is similar to lattices where for weak disorder $\ell_{\text{opt}} \sim r$, where r is the Euclidean distance [PSHB99]. In weak disorder, a path longer than the minimal distance is usually more costly. Thus, a weak disorder does not change the universality class of the length of the optimal path except in the case of "ultra small worlds" $2 < \gamma < 3$.

In contrast, a strong disorder dramatically changes the universality class of the optimal path. Theoretical considerations (Eqs. (12.2) and (12.3)) predict that in the case of WS and ER (Figure 12.3) and scale-free graphs with $\gamma > 4$, $\ell_{\text{opt}} = N^{1/3}$, whereas for scale-free graphs with $3 < \gamma < 4$, $\ell_{\text{opt}} \sim N^{(\gamma-3)/(\gamma-1)}$. Figure 12.5 shows the linear behavior of ℓ_{opt} versus $N^{1/3}$ for $\gamma \geq 4$. The quality of the linear fit becomes poor for $\gamma \to 4$. At this value, the logarithmic divergence of the second moment of the degree distribution occurs and one expects logarithmic corrections, i.e., $\ell_{\text{opt}} \sim N^{1/3} / \ln N$ [BWC+07]. Similar results were obtained for the scale-free networks in which the weights are associated with nodes rather than links [BWC+07]. The exact

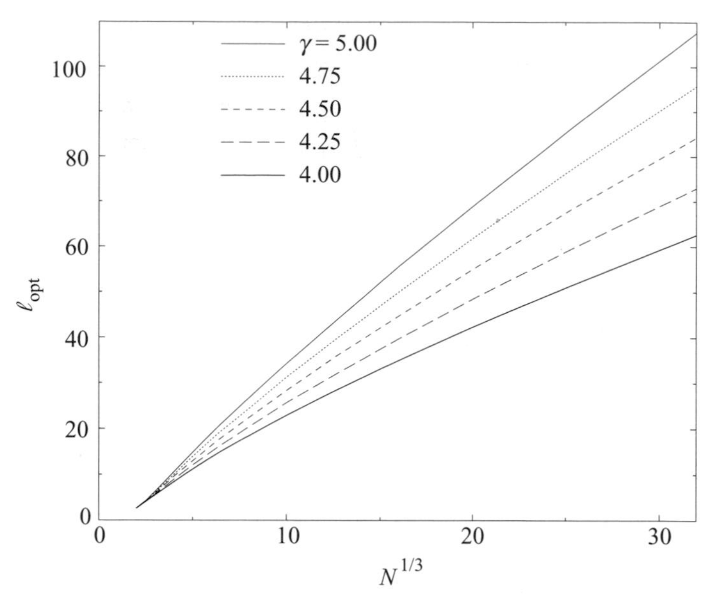

Figure 12.5 The dependence of ℓ_{opt} on $N^{1/3}$ for $\gamma \geq 4$.

nature of the percolation cluster at $\gamma < 3$ is not yet clear, since in this regime the transition does not occur at a finite concentration (see Chapter 10), and thus analytical results for optimal paths in strong disorder are still missing.

In summary, we have presented results for the optimal distance in ER, WS, and scale-free networks in the presence of strong and weak disorder. It was shown that in ER and WS networks for strong disorder, the optimal distance ℓ_{opt} scales as $N^{1/3}$. For the strong disorder limit in scale-free networks, theory and simulations show that ℓ_{opt} scales as $N^{1/3}$ for $\gamma > 4$ and as $N^{(\gamma-3)/(\gamma-1)}$ for $3 < \gamma < 4$. Thus, the optimal distance increases dramatically in strong disorder when it is compared to the known small-world result $\ell \sim \ln N$ and the "small-world" nature for these networks is destroyed. In contrast, simulations [BBC+03] also suggest that for $2 < \gamma < 3$, ℓ_{opt} scales as $\ln^{\gamma-1} N$, which is also much larger than the "ultra small-world" result $\ell \sim \ln \ln N$ [CH03], but can still be regarded as small world. The same scaling also applies to distances on the minimum spanning tree, which behave similarly to paths in strong disorder [DD01]. We also showed numerically that in weak disorder $\ell_{\text{opt}} \sim \ln N$ in all types of networks considered.

12.3 Correlated weighted networks

In the networks discussed above the weights of the links are selected randomly and independently from some distribution. No correlations exist between the node degrees and the weights of the links.

A study of the correlations between the degrees and the link weights is presented in [BBPV04]. In [BBPV04] two networks were studied, the world wide airport network (WWAN) and the network of scientific collaborations (SCN). In both networks, weights were defined, based on the number of seats in flights between airports each year (for the WWAN) and on the number of papers of coauthors, normalized by the number of authors of the paper (for the SCN).

For each node in such a weighted network one can define the total strength of a node i as

$$s_i = \sum_j w_{ij} \,, \tag{12.5}$$

where w_{ij} is the weight of the link between node i and node j. The strength of a node represents some measure of the node's functional importance in the network. One can define $s(k)$, the average strength of a node of degree k, as

$$s(k) = \frac{\sum_i s_i \delta_{k_i,k}}{\sum_i \delta_{k_i,k}} \,, \tag{12.6}$$

where $\delta_{k_i,k}$ is the Kronecker delta, giving one if the degree of node i is k and zero otherwise.

If weights are random and independent of the node properties then one expects to have

$$s(k) = \left\langle \sum_j w_{ij} \right\rangle = k \langle w \rangle \,. \tag{12.7}$$

However, for the networks studied in [BBPV04] it was found that for some networks there are correlations between node degrees and link weights. The dependence can be approximated by a power law

$$s(k) \sim k^{\beta} \,. \tag{12.8}$$

For uncorrelated networks one expects to find $\beta = 1$ in accordance with Eq. (12.7). This is indeed the case for the SCN. However, for the WWAN it was found [BBPV04] that $\beta = 1.5$. That is, larger airports, offering more destinations, also tend to have a larger fraction of high traffic lines between them.

One can also study the dependence of the weight of the link between two nodes on the degrees of the nodes. This, again, can be approximated by a power law [BBPV04]

$$\langle w_{ij} \rangle \sim (k_i k_j)^{\theta} \,, \tag{12.9}$$

for some θ. Even if correlations between the weight and degrees do exist, assuming no topological correlations between the degrees of connected nodes exist, it follows that

$$s_i \sim k_i \langle w_{ij} \rangle \sim k_i^{1+\theta} k_j^{\theta} \,. \tag{12.10}$$

This, in conjunction with Eq. (12.8), leads to $\beta = 1 + \theta$, which, indeed, is found to hold for the WWAN and the SCN networks. Further discussion on optimal distances in correlated weighted networks and the WWAN may also be found in [WBC$^+$06].

12.4 Summary

We presented results for optimal paths in networks with weighted links. For strong disorder, where the maximal weight along the path dominates the sum, we find that $\ell_{\text{opt}} \sim N^{1/3}$ in both Erdős–Rényi (ER) and Watts–Strogatz (WS) networks. Thus the small-world property of $d \sim \log N$ is no longer valid and the optimal distances are power law in N.

For scale-free networks, with degree distribution $P(k) \sim k^{-\gamma}$, we found that ℓ_{opt} scales as $N^{(\gamma-3)/(\gamma-1)}$ for $3 < \gamma < 4$ and as $N^{1/3}$ for $\gamma \geq 4$. Thus, for these networks, the small-world nature is destroyed. For $2 < \gamma < 3$, our numerical results suggest that ℓ_{opt} scales as $\ln^{\gamma-1} N$. We also found numerically that for weak disorder $\ell_{\text{opt}} \sim \ln N$ for both the ER and WS models as well as for scale-free networks [BBC$^+$03]. Thus, scale-free networks can be regarded as a generalization of ER graphs.

Exercises

12.1 What is the fractal dimension of a minimum spanning tree?

12.2 Simulate a minimum spanning tree and use the box covering method (Chaper 7) to calculate its fractal dimension.

12.3 Simulate a weighted network with a weak disorder and use the box counting method to find the crossover distance between the fractal and the exponentially growing part.

12.4 (Research) Study the (directed) optimal path length in directed networks with strong disorder (see Chapter 11 for the percolation properties of random directed networks).

PART III

NETWORK FUNCTION: DYNAMICS AND APPLICATIONS

13 Optimization of the network structure

In this chapter we present results for the robustness of complex networks under multiple waves of simultaneous targeted attacks in which the highest degree nodes are removed and random attacks (or failures), in which fractions p_t and p_r, respectively, of the nodes are removed until the network collapses [TPC+05]. It is shown that the network design, which optimizes network robustness, has a bimodal degree distribution, with a fraction r of the nodes having degree $k_2 = (\langle k \rangle - 1 + r)/r$ and the remainder of the nodes having degree $k_1 = 1$, where $\langle k \rangle$ is the average degree of all the nodes. The optimal value of r approaches p_t/p_r for $p_t/p_r \ll 1$.

13.1 Introduction

The resilience of real-world networks to random attacks or to attacks targeted at the highest degree nodes is of much interest [AJB00, CEbH00, CEbH01, CNSW00, Pax97, PSS05, VSS04]. Many real-world networks are robust to random attacks but vulnerable to targeted attacks on the most important nodes. It is important to understand how to design networks that are optimally robust against both types of attacks, with examples being terrorist attacks on physical networks and attacks by hackers on computer networks. In Chapter 10 we considered the case in which there was only one type of attack on a given network, that is, the network was subject to either a random attack or to a targeted attack but not subject to different types of attack simultaneously, which is a more realistic scenario. This scenario can be modeled as a sequence of "waves" of targeted and random attacks that remove fractions p_t and p_r of the original nodes, respectively. In the model presented here, the ratio p_t/p_r is kept constant while the individual fractions p_t and p_r approach zero. After some time (after m waves of random and targeted attacks), the network will become disconnected; at this point a fraction $f_c = m(p_t + p_r)$ of the nodes have been removed. This value f_c characterizes the robustness of the network. The larger the f_c, the more robust is the network. We describe here a mathematical approach [TPC+05] to study such simultaneous attacks and show the optimal network design, one that maximizes f_c. In the analysis, we compare the robustness of networks that have the same "cost" of construction and maintenance, where the cost is defined to be proportional to the average degree $\langle k \rangle$ in the network.

13.2 Optimization analysis

We study and compare mainly two types of random networks.

(i) *Scale-free networks*. As discussed in Chapter 3, many real-world computer, social, biological, and other types of networks have been found to be scale free, i.e., they exhibit degree distributions of the form $P(k) \sim k^{-\gamma}$. For large scale-free networks with exponent γ less than 3, for random attacks, essentially all nodes must be removed for the network to become disconnected [CEbH00, CNSW00]. On the other hand, because the scale-free distribution has a long power-law tail (i.e., hubs with large degree), these networks are very vulnerable to targeted attacks.

(ii) *Networks with bimodal degree distributions.* For resilience to single random or single targeted attacks, certain bimodal distributions are superior to any other network [PSS05, VSS04]. Here we investigate whether these networks are also the most resilient to simultaneous random and targeted attacks.

We present the following argument that suggests that the degree distribution that optimizes f_c is a bimodal distribution in which a fraction r of the nodes has degree

$$k_2 = \frac{\langle k \rangle - 1 + r}{r}, \tag{13.1}$$

and the remainder has degree $k_1 = 1$, and we show that r is of the order of p_t/p_r. To optimize against random removal, we maximize the quantity $\kappa \equiv \langle k^2 \rangle / \langle k \rangle$ (see Chapter 10), since for random removal the threshold is

$$q_c^{\text{rand}} = 1 - \frac{1}{\kappa - 1}. \tag{13.2}$$

Since we keep $\langle k \rangle$ fixed, κ is just the variance of the degree distribution and is maximized for a bimodal distribution in which the lower degree nodes have the smallest possible degree $k_1 = 1$ and the higher degree nodes have the highest possible degree consistent; when $\langle k \rangle$ is kept fixed, $k_2 = (\langle k \rangle - 1 + r)/r$. Thus, k_2 is maximized when r assumes its smallest possible value, $r = 1/N$. On the other hand, if all of the high-degree nodes are removed by targeted attacks, the network will be very vulnerable to random attack. Therefore, we want to delay as long as possible a situation in which all of the high-degree nodes are removed by targeted attacks – which argues for not choosing an r value as small as possible but instead choosing

r such that some high connectivity nodes remain as long as there are some low connectivity nodes. Such a condition is achieved when r is of the order of p_t/p_r.

The method we employ for determining the threshold makes use of the results obtained for random and targeted attacks (Chapter 10) and of the general condition for a random network to contain a giant component (see Chapter 9):

$$\kappa \equiv \frac{\langle k^2 \rangle}{\langle k \rangle} \geq 2. \tag{13.3}$$

Random removal of a fraction p_r of nodes from a network with degree distribution $P_0(k)$ results in a new degree distribution [CEbH00]

$$P(k) = \sum_{k_0=k}^{K} P_0(k) \binom{k_0}{k} (1 - p_r)^k \, p_r^{k_0-k}, \tag{13.4}$$

where K is the upper cutoff of the degree distribution. Targeted removal of a fraction p_t of the highest degree nodes reduces the value of upper cutoff K to $\tilde{K}$, which is implicitly determined by the equation

$$p_t = \sum_{k=\tilde{K}}^{K} P_0(k). \tag{13.5}$$

As shown in Chapter 10, the removal of high-degree nodes also modifies the degree distribution. This effect is equivalent to the random removal of a fraction of $\tilde{p}$ nodes where

$$\tilde{p} = \frac{\sum_{k=\tilde{K}}^{K} k P_0(k)}{\langle k \rangle_0}. \tag{13.6}$$

The average $\langle k \rangle_0$ has taken over the degree distribution before the removal of nodes [CEbH01]. Equation (13.4) with p_r replaced by $\tilde{p}$ can then be used to calculate the effect of removing the links. Starting with a certain initial degree distribution, we calculate $P(k)$ recursively by alternating between random and targeted attacks using Eqs. (13.4), (13.5), and (13.6), and calculate κ after each wave of attacks. When $\kappa < 2$ global connectivity is lost and $f_c = m(p_r + p_t)$, where m is the number of waves of attacks performed.

Numerical simulations [TPC+05] of Eqs. (13.3)–(13.6) for bimodal networks show that (a) for small values of p_t, p_r and p_t/p_r, the threshold f_c depends only on p_t/p_r, (b) only the scaled variable $r/(p_t/p_r)$ is relevant, and (c) the maximum values of f_c for various values of p_t/p_r are obtained when k_2 is maximum (i.e., when $k_1 = 1$, see Eq. (13.1)).

We are now in a position to determine the value of r, r_{opt}, that optimizes f_c. In Figure 13.1, we plot f_c as a function of the scaled parameter $r/(p_t/p_r)$ with k_2 set to the maximum value possible for each value of r. Note that there is a "transition" at a

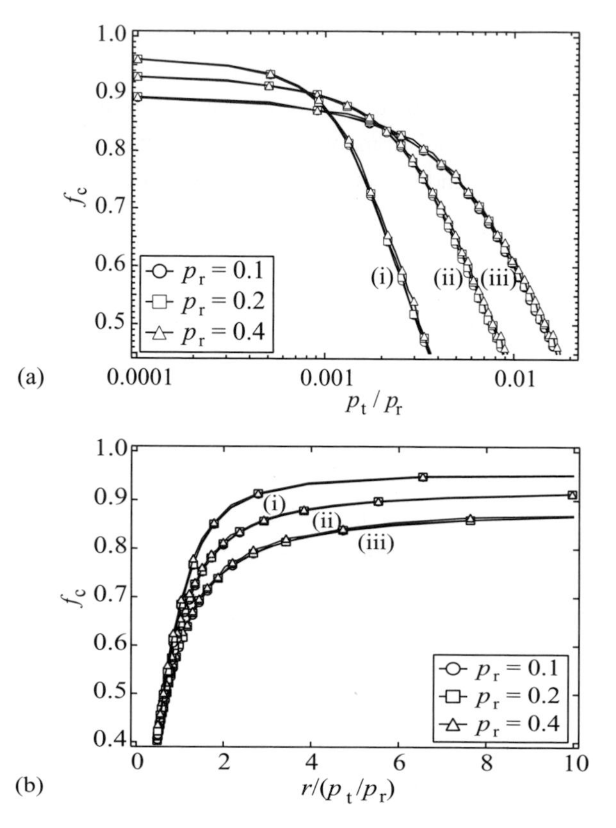

Figure 13.1 (a) The threshold f_c of three bimodal networks with $\langle k \rangle = 3$, with (i) $r = 2 \times 10^{-3}$ and $k_2 = 200$, (ii) $r = 5 \times 10^{-3}$ and $k_2 = 90$, and (iii) $r = 10^{-2}$ and $k_2 = 50$. The results are plotted as a function of the ratio p_t/p_r for three fixed values of p_r. These plots show that the values of the threshold are dependent only on the ratio p_t/p_r and are independent of the value of p_r itself. (b) Scaled plot of the data in (a). The data show that the plots collapse in the region where $r/(p_t/p_r) \lesssim 1$. After [TPC+05].

well-defined value of $r/(p_t/p_r)$ at which f_c increases rapidly to a shallow maximum f_c^{opt} at $r_{\text{opt}}/(p_t/p_r) \approx 1.7$. This value of $r_{\text{opt}}/(p_t/p_r)$ is valid for $p_t/p_r \ll 1$. In order to determine $r_{\text{opt}}/(p_t/p_r)$ over a wider range, extensive numerical calculations have been performed for $10^{-3} < p_t/p_r < 0.1$ (Figure 13.2). For each value of p_t/p_r, the value $r_{\text{opt}}/(p_t/p_r)$, where f_c takes its maximum value was calculated; it was found that

$$\frac{r_{\text{opt}}}{p_t/p_r} \approx 1.7 - 5.6\left(\frac{p_t}{p_r}\right) + \mathcal{O}\left(\frac{p_t}{p_r}\right)^2 \tag{13.7}$$

within the range of the calculations. For larger values of p_t/p_r, $r_{\text{opt}} = 1$ and from Eq. (13.1), all nodes have degree $\langle k \rangle$. In Figure 13.3, we plot the values of the optimal threshold f_c^{opt} by a thick solid curve.

Figure 13.3 also shows the values of the threshold f_c for the same bimodal network for fixed r as a function of p_t/p_r. As shown, these configurations are not significantly

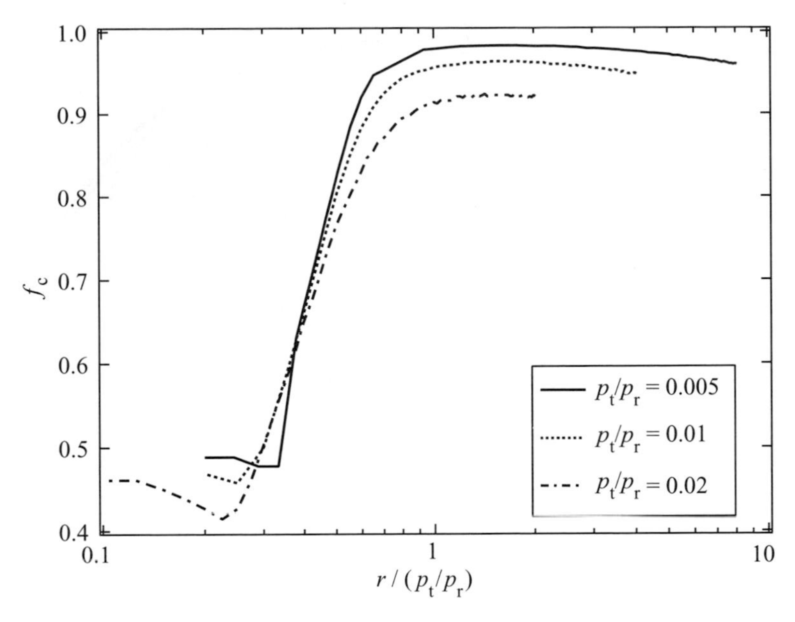

Figure 13.2 The threshold f_c versus the scaled parameter $r/(p_t/p_r)$ for a bimodal network with $\langle k \rangle = 3$ and k_2 maximum (i.e., $k_1 = 1$). After [TPC+05].

less robust than the optimal configuration. Thus, even if one does not know the ratio p_t/p_r exactly, one can design networks that will be relatively robust. For example, the bimodal network with $r = 0.03$ is relatively robust for $p_t/p_r \lesssim 0.1$ and the bimodal network with $r = 0.09$ is robust for $p_t/p_r \lesssim 1$. Also plotted in Figure 13.3 (using the same method) is the optimal scale-free network with $\langle k \rangle = 3$. We see that the optimal bimodal network is more robust than the optimal scale-free network and we can even pick a configuration with fixed r (e.g., $r = 0.03$) that is more robust than the optimal scale-free network in most ranges of p_t/p_r.

Figure 13.4 shows a typical optimal realization of a bimodal network. The network of $N = 100$ nodes consists of rN nodes with $k = k_2$ ("hubs") that are highly connected among themselves; the nodes of single degree are each connected to one of these hubs. Note that although the hubs are highly connected among themselves, they do not form a complete graph – every hub is not connected to every other hub. For larger N, the fraction of hubs to which a given hub connects decreases but the robustness of the network is unchanged.

13.3 General results

Given a random network of size N and degree distribution $P(k)$. Consider an attack where each site survives with probability Ae^{-ak}, where k is the degree of the site. A

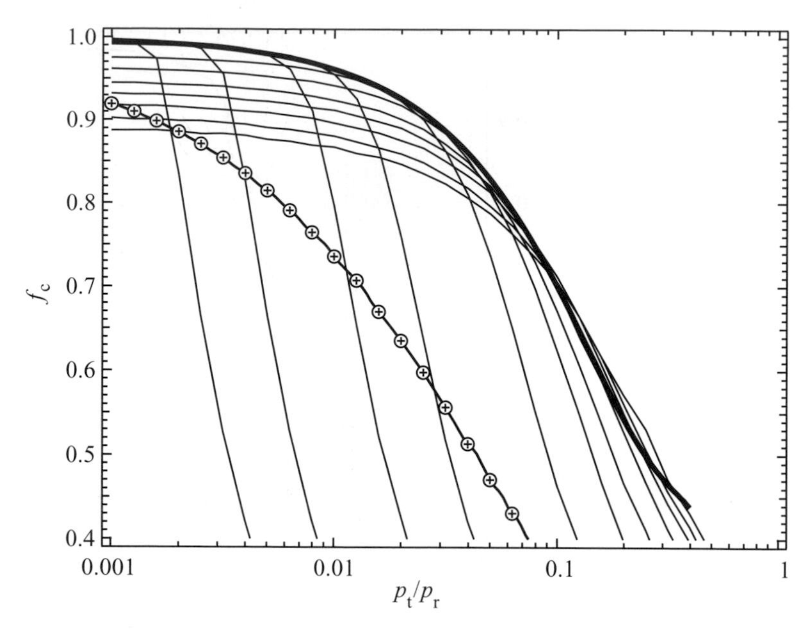

Figure 13.3 The threshold f_c versus p_t/p_r. The topmost (thickest) curve is for a bimodal network with $\langle k \rangle = 3$ with $k_1 = 1$ and with r optimized by Eq. (13.7) for each value of p_t/p_r. The values of the threshold for the same bimodal network with $k_1 = 1$ when we fix r independent of p_t/p_r are plotted in thin curves. The values of r are $r = 0.001, 0.002, 0.005, 0.01, 0.03, 0.05, 0.07, 0.09, 0.11, 0.13$, and 0.15, from left to right. The curve marked with crossed circles ($\oplus$) is a plot of the threshold values for a scale-free network with $\langle k \rangle = 3$, $N = 10^4$, and with an exponent chosen for each p_t/p_r to optimize the threshold. Note that the thresholds for bimodal networks with $0.03 \lesssim r \lesssim 0.09$ are always higher than those for the optimized scale-free networks. After [TPC+05].

is a parameter that determines the strength of the attack (i.e., the fraction of surviving sites as calculated later). The parameter a determines the preference for high-degree sites. For $a \to \infty$ the process becomes a targeted attack on the most highly connected sites. For $a \to 0$ the process becomes a random breakdown of sites. We will show that for all such attacks with $a > 0$, the optimal network configuration for resilience has a finite number of peaks in the degree distribution (for $a \leq 0$ this conclusion cannot be deduced since the degree of some nodes diverges and the distribution is discrete; therefore, the criterion for the threshold may not be applicable).

In the following, we determine the degree distribution of the most optimized uncorrelated random network having average degree $\langle k \rangle = c$. This requirement ensures that we compare networks having the same cost, i.e., the same total number of links.

The fraction of remaining sites at the percolation threshold is determined by the criterion:

$$p_c = \sum_{k=0}^{\infty} P(k) A_c e^{-ak} , \tag{13.8}$$

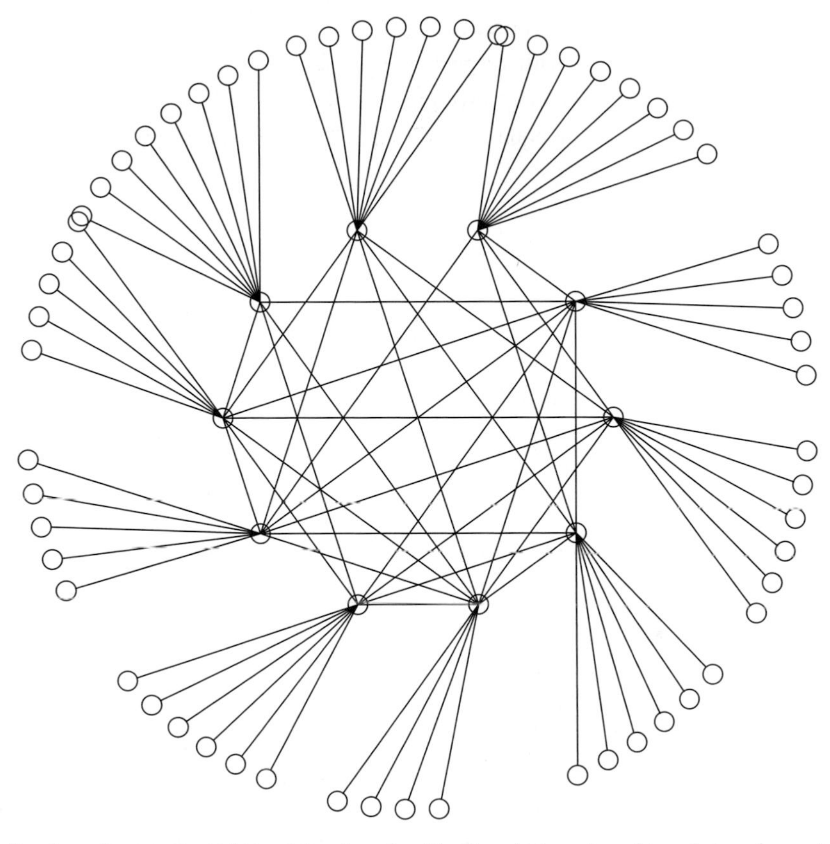

Figure 13.4 Realization of an optimal bimodal network with $N = 100$ nodes, $\langle k \rangle = 2.1$ and $r = 0.1$; thus, there are $rN = 10$ "hub" nodes of degree 12, as found from Eq. (13.1). After [TPC+05].

where $P(k)$ is the (original) degree distribution, and A_c is determined by the criterion for the existence of the giant component [CEbH00]:

$$\frac{1}{c}\sum_{k=0}^{\infty} k(k-1)P(k)A_c e^{-ak} = 1 . \tag{13.9}$$

There are two more requirements from the distribution: normalization

$$\sum_{k=0}^{\infty} P(k) = 1 , \tag{13.10}$$

and average degree

$$\sum_{k=0}^{\infty} kP(k) = c . \tag{13.11}$$

The goal is to find the distribution $P(k)$ minimizing p_c for a given average degree c and attack parameter a. To do so, we first have to define $g^2(k) \equiv P(k)$ (since otherwise negative values are also allowed and no extremum is found). We differentiate Eq. (13.8) with respect to $g(k)$ for each k and also with respect to A_c, and add the derivatives of Eqs. (13.9)–(13.11) with Lagrange multipliers (denoted as γ_1, γ_2, and γ_3). The equations obtained by differentiation with respect to $g(k)$ for some k are:

$$g(k)\left(A_c e^{-ak} + \gamma_1 k(k-1)A_c e^{-ak} + \gamma_2 + \gamma_3 k\right) = 0\,, \tag{13.12}$$

and by differentiating with respect to A_c we obtain,

$$\sum_{k=0}^{\infty}\left(P(k)e^{-ak} + \gamma_1 P(k)k(k-1)e^{-ak}\right) = 0\,. \tag{13.13}$$

From Eq. (13.12) it follows that for each k, either $g(k) = 0$ (and therefore also $P(k) = 0$) or otherwise,

$$A_c e^{-ak} + \gamma_1 k(k-1)A_c e^{-ak} = -\gamma_2 - \gamma_3 k\,. \tag{13.14}$$

Using the mean value theorem, it follows that between any two solutions of Eq. (13.14) there must be a value of k such that $e^{-ak}(1 + \gamma_1 k(k-1)) = \gamma_2/(A_c a)$. Now, according to Rolle's theorem [Apo67], between any two solutions of this equation there must be a point satisfying $e^{-ak}(1 + \gamma_1 k(k-1)) = 0$. The exponent is always positive; therefore, this equation has at most two solutions, and Eq. (13.14) has at most four solutions. The obtained optimal distribution has, therefore, at most four different degrees, and the rest are absent from the network. The requirement of the solutions to be positive integer values of k, and the global constraint in Eq. (13.13), may further decrease the number of possible degrees.

13.4 Summary

In summary, we have provided a qualitative argument and numerical results indicating that the most robust network against multiple waves of targeted and random attacks has a bimodal degree distribution with a fraction r of the nodes having degree $k_2 = (\langle k\rangle - 1 + r)/r$ and the remainder of the nodes having degree 1. The optimal value of r is approximately $1.7\,(p_t/p_r)$ for $p_t/p_r \ll 1$. For larger values of p_t/p_r, the optimal value of r is 1 and all nodes have degree $\langle k\rangle$. Even if p_t/p_r is not exactly known, a value of r can be chosen which maximizes the network robustness over a wide range of values of p_t/p_r, as seen in Figure 13.3.

Note that although the optimal distribution found here and that found in [PSS05] are both bimodal, the values of the parameters characterizing these distributions are different. As found in [PTHS04], a network with optimal resilience to either random or targeted attack has $r = 1/N$ and $k_2 \sim r^{-2/3}$. Finally, note that it is possible to prove analytically that for the case in which a single targeted attack followed by a single random attack results in the network becoming disconnected, the optimal distribution is also bimodal, with $k_1 = 1$, $k_2 = (\langle k \rangle - 1 + r)/r$ and r of the order of p_t/p_r [New02c], supporting the results found here for multiple waves of attacks.

Optimization of a network structure is not limited to optimization of the network robustness. Many other criteria for optimization may be used. Networks may be optimized for high throughput (see, e.g., [BF06]), for high tolerance to cascading failures (see, e.g., [ML02, Mot04]), and for many other criteria. For a review, see [MT07].

14 Epidemiological models

14.1 Introduction

The study of the spread of epidemics is based on the notion that a disease is conveyed by contact between an infected individual and an uninfected individual who is susceptible to the disease. An *endemic state* is reached if a finite fraction of the population is infected. A state where (almost) all the population is infected is called a *pandemic state*. Similarly, this notion may describe the spread of a computer virus through a network of computers.

In previous chapters we surveyed the subject of attacks on networks, by targeting individuals either randomly or intentionally. In particular, the Internet and the WWW were shown to be robust to random breakdowns and vulnerable to targeted attacks, due to their broad distribution of node degrees. However, these chapters focused on the results of damaging computers by an outside source, and did not take into account the possibility of propagation of a problem throughout the connections among individuals in a population or among computers in a network, by way of the spreading of a disease or a computer virus.

Several models have been proposed for epidemic dynamics, differing in the disease stages, the dynamical parameters, and the underlying structure of contacts (see [AM92] for a survey). The most common models for disease dynamics are the susceptible–infected–recovered (SIR), susceptible–infected–susceptible (SIS), and susceptible–infected–recovered–susceptible (SIRS) models, which represent the stages of the disease for each individual in a network. In this chapter we will focus on the SIR model, where an infected individual is infective for some period of time, and then recovers and can no longer infect or become infected by the disease, and on the SIS model [PV01b], where an individual is susceptible to the disease immediately after recovery.

Another issue regarding the model used is the underlying network topology. The network of (possible) contacts between individuals determines which individuals can infect each other. In general, this network may also be dynamic and change during the spread of the disease. We will assume that the network is static during the epidemic outbreak. We will also assume that the network is sparse, i.e., that the number of links is proportional to the number of individuals. We will focus on scale-free network topologies.

Moreover, in Chapter 15 we will also allow a fraction of the individuals to be immunized, i.e., these individuals cannot be infected. When a vaccination for a disease exists, immunizing certain individuals against being infected by a disease as a preemptive method may be the most efficient way to prevent loss of time and funds (and, of course, suffering, when dealing with infected people) as a result of the disease. Obviously, immunization of the entire population will entirely eradicate the disease, but this is not always possible, or may involve high costs and effort. Therefore, the choice of which individuals to immunize is an important step in the immunization process, and may increase the efficiency of an immunization strategy.

14.2 Epidemic dynamics and epidemiological models

A contagious disease may become an epidemic if the number of infected (sick) individuals is of the order of the number of individuals in the whole system. When there are no more sick individuals in the system (only susceptible left, and the others are removed), we can say that the disease is cured.

In many cases, an epidemic is indeed being contained. The main consideration therefore should be how much time is required for the system to reach such a stage, and how many individuals are being infected throughout the process. If both the length of time of the contagious period and the number of people exposed to it can be reduced, then the amount of suffering and loss of resources can be reduced as well.

Several models have been developed to describe the propagation of an epidemic disease in a population. In most models the population is divided into several groups according to their status, as presented below. A more detailed account of epidemiological models can be found in standard textbooks on epidemiology (see, for example, [AM92]).

Susceptible individuals are those who are not infected with the disease at the moment, and thus are not sick and are also not carriers of the disease. However, they are susceptible to the disease, i.e., they may be infected upon contact with infected individuals. Usually they are assumed only to develop the disease if they come in contact with an infected individual. Otherwise, they remain susceptible. In some models (depending on the duration of the outbreak and the population considered), susceptible individuals may be added to the population by birth, and removed by death from causes unrelated to the epidemic.

Infected individuals are individuals who carry the disease, and may infect other people with whom they come in contact. For the dynamics of the model, it does not really matter whether the individuals actually are sick or even whether they still carry the disease. The important point for the model dynamics is whether they are still infectious. Usually, the infectious stage of the disease is assumed to take some finite time. More exactly, the infected individuals are assumed to change at some rate back to the susceptible state or to a non-infectious state. For some diseases, which are not cured, and which do not convey high mortality rates, the infected individuals may stay in this state for the entire model duration.

Removed (or recovered) individuals are those individuals who are completely removed from the model in terms of the dynamics. This may be due to natural immunization to the disease, vaccination processes, death, or recovery from a disease, after which natural immunization is caused. Individuals in the removed state are assumed not to be susceptible to infections and not to be able to infect others.

14.2.1 The SIS model

In the SIS model each individual in the population belongs to one of two groups: **S**, susceptible individuals, or **I**, infected individuals. An infected individual may infect his susceptible neighbors at some rate, r, and may recover back to the susceptible state at some rate, q.

The simplest setup is one in which the population is fully connected (a complete graph). That is, each individual may infect everyone at the same rate. As is usually done in physics, to make the rates reasonable, they should be inversely proportional to the size of the population. Thus, writing S for the fraction of susceptible individuals and I for the fraction of infected ones, one obtains the mean-field equations,

$$\frac{\mathrm{d}S}{\mathrm{d}t} = qI - rIS\,, \tag{14.1}$$

$$\frac{\mathrm{d}I}{\mathrm{d}t} = rIS - qI\,, \tag{14.2}$$

$$S + I = 1\,, \tag{14.3}$$

where t is the time. These equations have a stable fixed point at $S = q/r$ (and thus $I = 1 - q/r$). Therefore, when $r > q$, an endemic state exists where a finite fraction of the population is infected at all times, whereas when $q > r$, an epidemic outburst will decay exponentially in time, and no stable endemic state exists. This is the first sign that epidemic outbursts are threshold phenomena, depending on a single parameter, $\varphi \equiv r/q$. For $\varphi < \varphi_c = 1$, no endemic state exists, whereas for $\varphi > \varphi_c$, an endemic state does exist.

The SIS model in networks

The behavior of the SIS model in scale-free networks was first considered in [PV01b]. Assume that the probability that a node is infected depends only on its degree. That is, no dependence on the node's surroundings exists (this is somewhat similar to effective medium approximations). Denote by ρ_k the fraction of nodes of degree k that are in the infected state at a given time. Since the important parameter is $\varphi \equiv r/q$, time may be rescaled such that the rate of recovery is 1. The equation for ρ_k thus reads,

$$\frac{\mathrm{d}\rho_k(t)}{\mathrm{d}t} = -\rho_k(t) + \varphi k(1-\rho_k(t))\Theta(\varphi)\,, \tag{14.4}$$

where $\Theta(\varphi)$ is the probability that a link leads to an infected node,

$$\Theta(\varphi) = \frac{1}{\langle k\rangle}\sum_k kP(k)\rho_k\,. \tag{14.5}$$

The stationary solution of this equation ($\mathrm{d}\rho_k(t)/\mathrm{d}t = 0$) is

$$\rho_k = \frac{k\varphi\Theta(\varphi)}{1+k\varphi\Theta(\varphi)}\,. \tag{14.6}$$

Equations (14.5) and (14.6) can now be solved simultaneously to obtain the total fraction of infected individuals in the stationary state:

$$\rho = \sum_k P(k)\rho_k\,. \tag{14.7}$$

For a regular graph, i.e., a constant number of links per node, where $P(k) = \delta_{k,k_0}$, the solution of the above equations yields $\Theta = 1 - (1/k_0\varphi)$ $(k_0\varphi > 1)$, otherwise $\Theta = 0$. In fact, the solution $\Theta = 0$ always exists, representing a state where no one is infected; however, this solution is only stable when $k_0\varphi < 1$. Thus, the critical threshold for the endemic state in this case is $\varphi_c = k_0^{-1}$.

In the case of a scale-free distribution, $P(k) = ck^{-\gamma}$, the expressions become complicated [PV01b]. The sum can be approximated by an integral and yields the hypergeometric function,

$$\Theta = F[1, \gamma-2, \gamma-1, -(m\varphi\Theta)^{-1}]\,. \tag{14.8}$$

In the special case where $\gamma = 3$, the integrals take a somewhat simpler form, leading to the approximate solution,

$$\rho \sim \mathrm{e}^{-1/(m\varphi)}\,. \tag{14.9}$$

Note that for all $\varphi > 0$ values, ρ is finite. Therefore, no finite threshold exists, and an endemic state always exists, regardless of how low the infection rate is. This is also true for all $\gamma < 3$ values, where the second moment diverges [PV01b], similarly to what was seen before in Chapter 10. In cases where the second moment is convergent, since the denominator in Eq. (14.6) is always larger than 1, it follows that $\rho_k \leq k\varphi\Theta(\varphi)$. Using this with Eq. (14.5) leads to $\Theta \leq \langle k \rangle^{-1} \sum_k k^2 P(k)\varphi\Theta$. Thus, whenever $\sum_k k^2 P(k)\varphi < \langle k \rangle$, the only possible solution is $\Theta = 0$ and no endemic state exists. We can conclude that for random networks with a converging second moment of the degree distribution, a finite threshold $\varphi_c \leq \langle k \rangle / \langle k^2 \rangle$ always exists [PV01b].

When considering a real-world application of these results, one must note that no threshold existing is only applicable in the limit $N \to \infty$. Since populations are finite and even rather small compared with thermodynamic standards, a cutoff and therefore a threshold will exist for every epidemic and it may even be reasonably large. Note that in Eq. (14.9) the size of the infected population decreases exponentially with $1/\varphi$. Thus, when $\varphi \approx 1/\ln N$, which is a reasonably high threshold, the epidemic will infect a finite number of individuals before dying out owing to stochastic effects.

14.2.2 The SIR model

The SIR model represents the development of a disease in a network of connected individuals. **S** stands for the susceptible stage, where the individual is healthy. **I** stands for the infected stage, where the individual is infected with the disease and can infect other individuals in contact with him. **R** is the removed stage, where the individual has either recovered and has acquired immunization to the disease or is otherwise permanently removed from the system.

In the numerical simulations by [MKC$^+$04], all the individuals (nodes) are at first susceptible, i.e., they are all healthy and none of them is immune to the disease. One node, chosen randomly, is infected. At each time step, every susceptible neighbor of an infected node has a probability of becoming infected itself, and each infected node has a probability of being removed from the system. It is assumed that both probabilities (infection and removal) are the same for each node and its neighbors (networks having different probabilities for each node are studied in [New02c]).

The SIR model as bond percolation

One of the nicest features of the SIR model is that despite being a dynamic model, it can be mapped into a completely static one [Gra83, New02c, WSS02a]. Consider a network where each node transmits the epidemic to each of its neighbors with rate r, and is removed after recovery time τ. The infection can therefore be considered as a Poisson process, with average $r\tau$. Thus, the probability of each neighbor *not* being infected is $e^{-r\tau}$.

The outcome of this process is therefore the same as bond percolation, in which each directed link is occupied with probability $p_b = 1 - e^{-r\tau}$. If r and τ have the same value for each node, the network can be considered as non-directed. Although information on the epidemic dynamics is lost by this description, the critical threshold for the dynamic model can be deduced from the bond percolation problem. Furthermore, the total fraction of infected individuals in the endemic state is the same as the size of the giant component of the percolation model, and the probability of a single disease event decaying before reaching the endemic state equals the fraction of finite components in the percolation networks. If τ is chosen from a distribution, a more involved site–bond percolation model is needed to find the exact threshold [KR07, Mil07].

To find the threshold for bond percolation in networks, one should consider the average number of links outgoing from a node that is, in itself, reached by following a link. This is similar to the course of the epidemic. If an infected individual infects, on average, at least one other individual, then the epidemic can reach an endemic state. Since a node can be reached by one of its k links, its probability of being reached is $kP(k)/(N\langle k\rangle)$, where N is the number of nodes, $P(k)$ is the fraction of nodes having degree (number of links) k, and $\langle k\rangle = \sum_k kP(k)$ denotes the average degree of nodes in the network. The probability of each of its $k-1$ outgoing links infecting its neighbor is p_b. Since the network is randomly connected, as long as the epidemic is not yet spread, the average number of influenced neighbors is:

$$\langle n_i\rangle = p_b \sum_k \frac{P(k)k(k-1)}{\langle k\rangle} . \tag{14.10}$$

Therefore, an endemic state can be reached only if $\langle n_i\rangle > 1$, leading to [CEbH00, CNSW00]

$$\left(\frac{\langle k^2\rangle}{\langle k\rangle} - 1\right) > p_b^{-1} . \tag{14.11}$$

From this expression, it can easily be seen that scale-free networks, with degree distribution $P(k) \sim k^{-\gamma}$, with $\gamma \leq 3$, having a divergent second moment, undergo the transition only at $p_b \to 0$ [CEbH00, CNSW00]. That is, an epidemic can spread in this network regardless of how small the infection probability is and how quick the recovery process is [PV01b].

14.2.3 The SIRS model

The last model we will discuss is the SIRS model. This model is a generalization of both the SIS and SIR models. In this model a susceptible individual can become infected by an infected neighbor. Infected individuals recover at some rate r, for some duration of time in which they are naturally immune and cannot be infected again, and with some rate q, they become susceptible again. This model reduces to the SIS model when $q \to \infty$ and to the SIR model when $q \to 0$. Discussing further the properties of the SIRS model is beyond the scope of this book, details can be found in, for example, [AM92].

Exercises

14.1 Solve the SIS model for the complete (fully connected) network using a master equation for the number of infected individuals.

14.2
(a) What is the probability that, starting at some arbitrary time, all infected nodes will recover before any node is infected?
(b) What is the probability that as $t \to \infty$ at least one event such as the one presented in (a) will occur?
(c) What is the probability of the population recovering?
(d) Can you explain the apparent discrepancy between this result and the behavior of the SIS model as discussed in this chapter?

14.3 Solve the SIR model for the complete graph.

14.4 Solve the SIRS model for the complete graph.

14.5 What is the infection threshold for a random regular graph of constant degree k:
(a) for the SIS model?
(b) for the SIR model?
(c) Solve the SIR model for a random network with bimodal degree distribution where a fraction p of the nodes have degree k_1, and the rest have degree k_2.
(d) Solve the SIS model for a bimodal network with a fraction p of the nodes having degree k_1 and the rest having degree k_2.
(e) Solve the SIR model for a bimodal network with a fraction p of the nodes having degree k_1 and the rest having degree k_2.

15 Immunization

In general, immunization can be seen as a site percolation problem. Each immunized individual can be regarded as a node that is removed from the network. The goal of the immunization process is to pass (or at least approach) the percolation threshold, leading to minimization of the number of infected individuals. The complete model of SIR and immunization can be considered as a site–bond percolation model, and the immunization is considered successful if the network is below the percolation threshold.

It is well established that immunization of randomly selected individuals requires immunizing a very large fraction, f, of the population, in order to arrest epidemics that spread upon contact with an infected individual [AJB00, AM92, CEbH00, HA87, MA84, PV01b, WY84]. Many diseases require 80%–100% immunization. For example, measles requires 95% of the population to be immunized [AM92]. The same is true for the Internet, where stopping computer viruses requires almost 100% immunization [AJB00, CEbH00, KWC93, PV01b]. On the other hand, targeted immunization of the most highly connected individuals [AJB00, AM92, CEbH01, CNSW00, LM01, PV02], while effective, requires global information about the network in question, rendering it impractical in many cases.

In this chapter we present a mathematical model and propose an effective strategy, based on the immunization of a small fraction of *random acquaintances* of randomly selected nodes [CHb03]. In this way, the most highly connected nodes are immunized, and the process prevents epidemics with a small finite immunization threshold and without requiring specific knowledge of the network. Other, efficient, immunization strategies which require global information include the adaptive high degree or centrality attacks [GLA+07, Hol04], and the equal graph partitioning method [CPH+08, PCS+07].

15.1 Random immunization

As discussed earlier, social networks are known to possess a broad degree distribution. Some examples are the web of sexual contacts [LEA+01], movie-actor networks, science citations, and cooperation networks [BJN+02, NWS02]. Computer networks, both physical (such as the Internet [YJB02]) and logical (such as the WWW [AJB99], email [EMB02] and trust networks [GGA+02]) are also known to possess wide, scale-free, distributions. As seen in Chapter 10, percolation theory on broad-scale networks shows that a large fraction f_c of the nodes need to be removed (immunized) before the integrity of the network is compromised. This is particularly true for scale-free networks, $P(k) = ck^{-\gamma}$ ($k \geq m$), where $2 < \gamma < 3$ – the case for many known networks [AB02, DM02, Str00] – where the percolation threshold $f_c \to 1$, and the network remains connected (contagious) even after removal of most of its nodes [CEbH00]. In other words, with a random immunization strategy almost all of the nodes need to be immunized before an epidemic is arrested (see Figure 15.1 below).

To calculate the immunization threshold, one should consider the site–bond percolation model. The considerations are the same as for the epidemic threshold (Eq. (14.11)), with the exception that a node may also be immunized, in which case it cannot propagate the disease. This adds another factor of $p_s = 1 - f$, the probability that a node is not immunized, to the calculation, leading to:

$$\left(\frac{\langle k^2 \rangle}{\langle k \rangle} - 1\right) > (p_s p_b)^{-1} . \tag{15.1}$$

As can be seen, in this case as well, the epidemic will only be arrested if $p_b p_s \to 0$, meaning that for every epidemic almost the entire population must be immunized in order to prevent the epidemic from spreading.

15.2 Targeted immunization: choosing the right people to immunize

When the most highly connected nodes are targeted first, removal of just a small fraction of the nodes results in disintegration of the network [AJB00, CEbH01, CNSW00]. This has led to the suggestion of targeted immunization of the hubs (the most highly connected nodes in the network) [DB02, PV02].[1]

The simplest targeted immunization strategy calls for the immunization of the most highly connected individuals. To use this approach, the number of connections

[1] It should be mentioned that applying high-degree adaptive immunization (and high centrality adaptive immunization), where the remaining degree (or centrality) of each node is reevaluated after every node removal, is found to be even more efficient [Hol04].

of each individual should be known (at least approximately, see [DB02]). In this case, the probability that a node is not immunized, when the immunization rate is f, is $\theta_f(k)$, where

$$\theta_f(k) = \begin{cases} 1 & k < k^* \\ c & k = k^* \\ 0 & k > k^*, \end{cases} \tag{15.2}$$

and k^* and $0 < c \leq 1$ are determined by the condition

$$\sum_k P(k)\theta_f(k) = 1 - f\,. \tag{15.3}$$

To find the critical immunization threshold using this strategy, one can again find the fraction giving, on average, one outgoing infective link per infected individual. This amounts to the requirement:

$$\sum_k \frac{k(k-1)P(k)\theta_{f_c}(k)}{\langle k \rangle} = p_b^{-1}\,. \tag{15.4}$$

Solving Eq. (15.4) in conjunction with Eq. (15.3) allows one to calculate the exact immunization threshold. The implications of partial knowledge of the node degrees, leading to functions other than $\theta_f(k)$ were studied in [DB02].

15.3 Acquaintance immunization: choosing the right people with minimal information

15.3.1 Description

One problem with the targeted immunization approach is that it requires a complete, or at least fairly good, knowledge of the degree of each node in the network. Such global information often proves hard to gather, and may not even be well defined (as in social networks, where the number of social relations depends on subjective judgements). The acquaintance immunization strategy proposed herein works at low immunization rates, f, and obviates the need for global information.

In this approach [CHb03], we choose a random fraction p of the N nodes and look for a random acquaintance with whom they are in contact (thus, the strategy is purely local, requiring minimal information about randomly selected nodes and their immediate environments). The acquaintances, rather than the originally chosen nodes, are the ones immunized. The fraction p may be larger than 1, because a node might be queried more than once, on average, whereas the fraction of nodes immunized f is always less than or equal to 1.

15.3.2 Analysis

Suppose we apply the acquaintance strategy on a random fraction p of the network. The critical fractions, p_c and f_c, needed to stop the epidemic can be calculated analytically. In each event, the probability that a particular node with k contacts is selected for immunization is $kP(k)/(N\langle k\rangle)$ [CEbH00, CNSW00]. This quantifies the known fact that randomly selected acquaintances have, on average, a higher degree than randomly selected nodes [Fel81, New03a].

Suppose we follow some possible branch in the course of the epidemic, starting from a random link of the spanning cluster. That is, we study the possible spread of the epidemic by considering nodes that are not immunized, and therefore are susceptible to the epidemic and may become infected. In some layer (hop distance from the starting point), l, we have $n_l(k')$ nodes of degree k'. In the next layer $(l+1)$, each of those nodes has $k'-1$ new neighbors (excluding the one through which we arrived). Let us denote the case where a node of degree k is susceptible to the disease (not immunized, and therefore it may be infected through the course of epidemic spreading) by s_k. To determine the number of nodes, $n_{l+1}(k)$, of degree k that are susceptible and are reached in the course of the epidemic, we multiply the number of links going out of the lth layer by the probability of reaching a node of degree k through following a link from a susceptible node, $p(k|k', s_{k'})$. Then, we multiply by the probability that this node is also susceptible, given both the node's and the neighbor's degrees, and the fact that the neighbor is also susceptible, $p(s_k|k, k', s_{k'})$. Since below and at the critical percolation, threshold loops are irrelevant [CEbH00], one can ignore them in calculating the threshold. Therefore,

$$n_{l+1}(k) = p_b \sum_{k'} n_l(k')(k'-1)p(k|k', s_{k'})p(s_k|k, k', s_{k'}) \,. \tag{15.5}$$

By using Bayes' rule:

$$p(k|k', s_{k'}) = \frac{p(s_{k'}|k, k')p(k|k')}{p(s_{k'}|k')} \,. \tag{15.6}$$

Assuming that the network is uncorrelated (no degree-degree correlations), the probability $\phi(k)$ of reaching a node with degree k via a link is independent of k':

$$\phi(k) \equiv p(k|k') = kP(k)/\langle k\rangle \,. \tag{15.7}$$

(A study of cases where correlations exist can be found in [BP02, BPV03, New02c].)

A random node (of degree k') is selected in each step with probability $1/N$. The probability of being redirected to a specific acquaintance is $1/k'$. Thus, the probability that the acquaintance is *not* selected in one particular attempt is $1 - 1/(Nk')$, and in all Np vaccination attempts, it is

$$\nu_p(k') \equiv \left(1 - \frac{1}{Nk'}\right)^{Np} \approx \mathrm{e}^{-p/k'} . \tag{15.8}$$

If the neighbor's degree is not known, the probability is $\nu_p \equiv \langle \nu_p(k) \rangle$, where the average (and all averages henceforth) is taken with respect to the probability distribution $\phi(k)$. In general, the probability that a node with degree k is susceptible is $p(s_k|k) = \langle \mathrm{e}^{-p/k} \rangle^k$, if no other information exists on its neighbors. If the degree of one neighbor (which is the one through which the epidemic propagated) is known to be k': $p(s_k|k, k') = \mathrm{e}^{-p/k'} \times \langle \mathrm{e}^{-p/k} \rangle^{k-1}$. Since the fact that a neighbor with a known degree is immunized does not provide any further information about a node's probability of immunization, it follows that $p(s_k|k, k') = p(s_k|k, k', s_{k'})$. Using the above equations, one obtains:

$$p(k'|k, s_k) = \frac{\phi(k')\mathrm{e}^{-p/k'}}{\langle \mathrm{e}^{-p/k} \rangle} . \tag{15.9}$$

Substituting these results in (15.5) yields:

$$n_{l+1}(k) = p_{\mathrm{b}} \nu_p^{k-2} \phi(k) \mathrm{e}^{-p/k} \sum_{k'} n_l(k')(k' - 1)\mathrm{e}^{-p/k'} . \tag{15.10}$$

Since the sum in (15.10) does not depend on k', it leads to the stable distribution of degree in a layer l: $n_l(k) = a_l \nu_p^{k-2} \phi(k) \mathrm{e}^{-p/k}$, for some a_l. Substituting this into (15.10) yields:

$$n_{l+1}(k) = n_l(k) p_{\mathrm{b}} \sum_{k'} \phi(k')(k' - 1) \nu_p^{k'-2} \mathrm{e}^{-2p/k'} . \tag{15.11}$$

Therefore, if the sum is larger than 1, the branching process will continue forever (the percolating phase), whereas if it is smaller than 1, immunization is subcritical and the epidemic is arrested. Thus, we obtain a relation for p_{c}:

$$\sum_k \phi(k)(k - 1) \nu_{p_{\mathrm{c}}}^{k-2} \mathrm{e}^{-2p_{\mathrm{c}}/k} = p_{\mathrm{b}}^{-1} , \tag{15.12}$$

where the case $p_{\mathrm{b}} \to 1$ corresponds to full immunization, i.e., stopping the epidemic regardless of its infection rate.

The fraction of immunized nodes is easily obtained from the fraction of nodes that is not susceptible,

$$f_{\mathrm{c}} = 1 - \sum_k P(k) p(s_k|k) = 1 - \sum_k P(k) \nu_{p_{\mathrm{c}}}^k , \tag{15.13}$$

where $P(k)$ is the regular distribution, and p_c is found numerically using Eq. (15.12). The term $\nu_{p_c}^{k-2}$ in Eq. (15.12) induces an exponential cutoff on the degree distribution of susceptible nodes for $0 < \nu_{p_c} < 1$. Therefore, the sum in Eq. (15.12) becomes finite for some finite $\nu_{p_c} > 0$. Substituting this into Eq. (15.13) indicates that $f_c \neq 1$, and is finite even in the thermodynamic limit.

A related immunization strategy calls for the immunization of acquaintances (AcI strategy) referred to by at least n nodes. (Above, we applied the special case $n = 1$.) The threshold is lower as n increases, and may justify, under certain circumstances, this somewhat more involved protocol.

To analyze the threshold for the double acquaintance ($n = 2$) case, we should replace the probabilities for susceptibility with the appropriate probabilities, considering the fact that a node is immunized only if 2 of its contacts point at it. Since the process is a Poisson process (in the limit of large N), the probabilities are:

$$p(s_k|k, k') = \mathrm{e}^{-p/k'} \left\langle \mathrm{e}^{-p/k} \right\rangle^{k-2} \times \left[\left\langle \frac{p\mathrm{e}^{-p/k}}{k} \right\rangle (k-1) + \left\langle \mathrm{e}^{-p/k} \right\rangle \left(1 + \frac{p}{k'}\right) \right], \tag{15.14}$$

and

$$p(s_k|k) = \left\langle \mathrm{e}^{-p/k} \right\rangle^{k-1} \left[\left\langle \frac{p\mathrm{e}^{-p/k}}{k} \right\rangle k + \left\langle \mathrm{e}^{-p/k} \right\rangle \right]. \tag{15.15}$$

We will use the notation $\nu_p \equiv \left\langle \mathrm{e}^{-p/k} \right\rangle$ and $\mu_p \equiv p \left\langle \mathrm{e}^{-p/k}/k \right\rangle$. Using Bayes' rule as before and substituting into Eq. (15.5), one obtains

$$n_{l+1}(k) = p_b \sum_{k'} n_l(k')\phi(k)\mathrm{e}^{-p/k}\mathrm{e}^{-p/k'} {\nu_p}^{k-3} \times \frac{(k'-1)\left[(k'-1)\mu_p + \left(1 + \frac{p}{k}\right)\nu_p\right]\left[(k-1)\mu_p + \left(1 + \frac{p}{k'}\right)\nu_p\right]}{\nu_p + k'\mu_p}. \tag{15.16}$$

It can now be seen that the kernel of Eq. (15.16) is separable into three functions:

$$n_l(k) = \phi(k){\nu_p}^{k-3}\mathrm{e}^{-p/k}\left(a_l + b_l k + \frac{c_l}{k}\right). \tag{15.17}$$

Substituting this back into Eq. (15.16) leads to the matrix notation

$$\begin{pmatrix} a_{l+1} \\ b_{l+1} \\ c_{l+1} \end{pmatrix} = p_b \sum_{k'} \frac{\phi(k')(k'-1)\mathrm{e}^{-2p/k'}{\nu_p}^{k'-3}}{\nu_p + k'\mu_p} \mathbf{M} \begin{pmatrix} a_l \\ b_l \\ c_l \end{pmatrix}, \tag{15.18}$$

where $\mathbf{M}$ is the matrix:

$$\mathbf{M} = \begin{pmatrix} A_p(k') & k'A_p(k') & \frac{A_p(k')}{k'} \\ \mu_p B_p(k') & k'\mu_p B_p(k') & \frac{\mu_p B_p(k')}{k'} \\ p\nu_p C_p(k') & k'p\nu_p C_p(k') & \frac{p\nu_p C_p(k')}{k'} \end{pmatrix}, \tag{15.19}$$

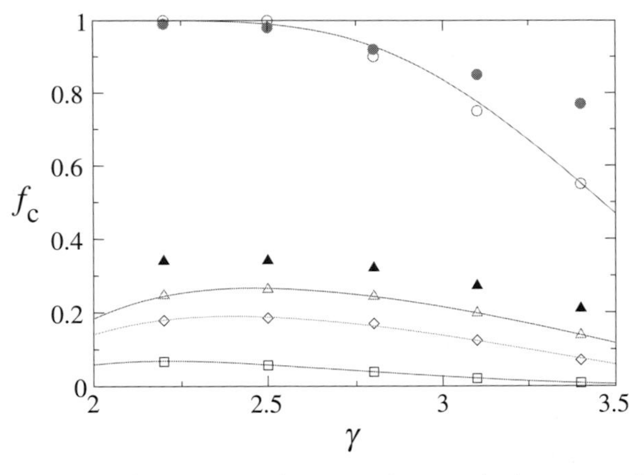

Figure 15.1 Critical immunization threshold, f_c, as a function of γ in scale-free networks (with $m = 1$), for the random immunization (○), acquaintance immunization (△), double acquaintance immunization (◇), and targeted immunization (□) strategies. Curves represent analytical results, whereas data points represent simulation data, for a population $N = 10^6$ (due to the finite size of the population, $f_c < 1$ for random immunization even when $\gamma < 3$). Full symbols are for random and acquaintance immunization of assortatively mixed networks (where links between nodes of degree k_1 and $k_2 (> k_1)$ are rejected with probability $0.7\,(1 - k_1/k_2)$). After [CHb03].

and we have used the definitions $B_p(k') \equiv \nu_p + k'\mu_p - \mu_p$, $C_p(k') \equiv \nu_p - \mu_p + \nu_p p/k'$ and $A_p(k') \equiv C_p(k')B_p(k') + p\mu_p\nu_p$.

Since this is a branching process, it is controlled by the largest eigenvalue of the matrix $\mathbf{N}$,

$$\mathbf{N} = \sum_{k'} \frac{\phi(k')(k'-1)\mathrm{e}^{-2p/k'}{\nu_p}^{k'-3}}{\nu_p + k'\mu_p}\mathbf{M}\,. \tag{15.20}$$

This eigenvalue can be calculated numerically using standard methods and the immunization threshold is obtained when $\gamma_1 \equiv \max_{\mathbf{v}} ||\mathbf{Mv}||/||\mathbf{v}||$, the largest eigenvalue of $\mathbf{N}$, satisfies $\gamma_1 = 1/p_{\mathrm{b}}$. This can be solved numerically for a given degree distribution $P(k)$. The critical value p_{c} is then obtained and can be used to evaluate f_{c}, the fraction of immunized individuals,

$$f_{\mathrm{c}} = 1 - \sum_k P(k){\nu_{p_{\mathrm{c}}}}^{k-1}(\nu_{p_{\mathrm{c}}} + pk\mu_{p_{\mathrm{c}}})\,. \tag{15.21}$$

15.3.3 Discussion

Figure 15.1 presents the immunization threshold f_{c} needed to stop an epidemic in scale-free networks with $2 < \gamma < 3.5$ (this covers most known cases). Plotted are curves for the (inefficient) random strategy, and the strategy advanced here, for the

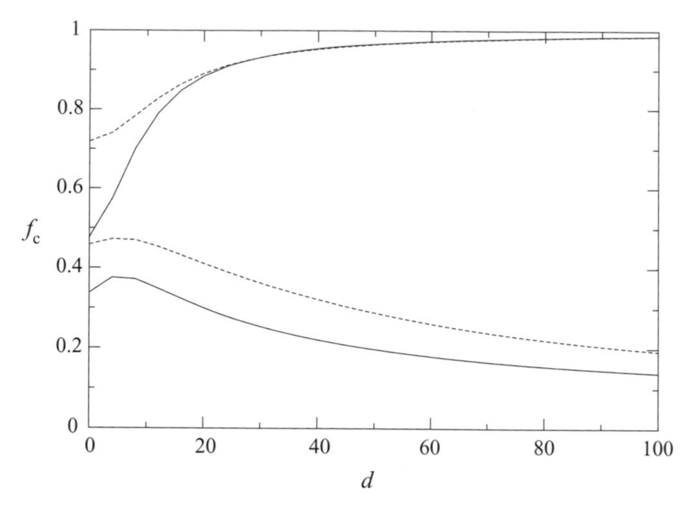

Figure 15.2 Critical concentration, f_c, for the bimodal distribution (of two Gaussians) as a function of d, the distance between the mean degrees of the high-degree nodes and low-degree nodes. The first Gaussian is centered at $k = 3$ and the second at $k = d + 3$ with height 5% of the first. Both have variance 2 (solid lines) or 8 (dashed lines). The top 2 lines are for random immunization. The bottom 2 lines are for acquaintance immunization. All curves are derived analytically from Eqs. (15.13) and (15.12). Note that also for the case $d = 0$, i.e., a single Gaussian, the value of f_c reduces considerably following the acquaintance immunization strategy. Thus, the strategy yields improved performance even for relatively narrow distributions [ASBS00]. After [CHb03].

cases of $n = 1$ and 2. Note that whereas $f_c = 1$ for networks with $2 < \gamma < 3$ (e.g., for the Internet), it decreases dramatically to values $f_c \approx 0.25$ with the AcI strategy. The figure also shows the strategy's effectiveness in the case of assortatively mixed networks [New02a], i.e., in cases where $p(k'|k)$ does depend on k, and high-degree nodes tend to connect to other high-degree nodes, which is the case for many real networks.

Figure 15.2 presents similar results for a bimodal distribution (consisting of two Gaussians) where high-degree nodes are rare compared with low-degree ones. This distribution is also believed to exist for some social networks, in particular for some networks of sexual contacts. The improvement gained by the use of the AcI strategy is evident in Figure 15.2.

Figure 15.3 presents the dependence of the immunization threshold f_c on the infection rate r in the SIR model.

These figures show that the acquaintance immunization (AcI) strategy is effective for any broad-scale distributed network. Results are presented for scale-free and bimodal distributions, which are common in many natural networks, as well as assortatively mixed networks (where high-degree nodes tend to connect to other high-degree nodes [New02a]).

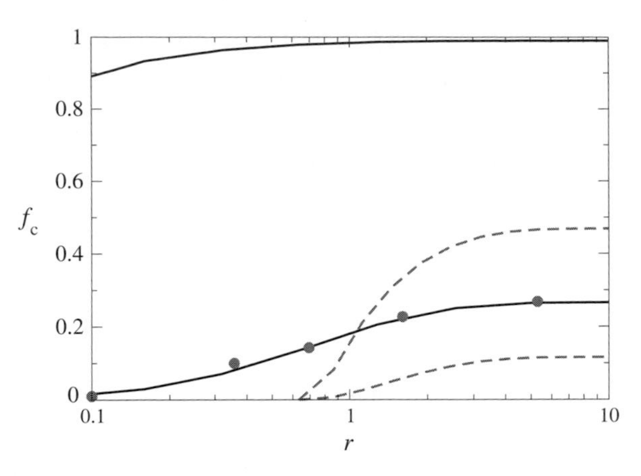

Figure 15.3

Critical concentration, f_c, versus r, the infection rate, for the SIR model with $\tau = 1$. The solid lines are for random (top) and acquaintance immunization (bottom) for scale-free networks with $\gamma = 2.5$. The dashed lines are for $\gamma = 3.5$ (top random, bottom acquaintance immunization). The full circles represent simulation results for acquaintance immunization for scale-free networks with $\gamma = 2.5$. After [CHb03].

15.3.4 Practical issues

Various immunization strategies have been proposed, mainly for the case of an already spread disease, which are based on tracing the chain of infection towards the super-spreaders of the disease [WY84]. This approach differs from the AcI approach, since it is mainly aimed at stopping an epidemic after the outbreak has begun. It is also applicable in cases where no immunization exists and only treatment for individuals already infected is possible. The AcI strategy, on the other hand, can be used even before the epidemic starts spreading, since it does not require any knowledge of the chain of infection.

In practice, any population immunization strategy must take into account issues of attempted manipulation. We would expect the AcI strategy to be less sensitive to manipulations than targeted immunization strategies. This is due to its dependence on acquaintance reports rather than on *self*-estimates of the number of contacts. Since a node's reported contacts pose a direct threat to the node (and relations), manipulations would probably be less frequent. Furthermore, some problems may be solved by adding some randomness to the process: for example, reported acquaintances are not immunized, with some small probability (smaller than the random epidemic threshold), whereas randomly selected individuals are immunized directly, with some low probability. This will have a small impact on the efficiency, while enhancing privacy and rendering manipulations less practical.

15.4 Numerical results for the SIR model

Figure 15.4 presents the dynamics of the spread of the disease in the SIR model, compared with the spreading dynamics after the implementation of the three methods of immunization that were discussed above (random, acquaintance, and targeted).

In the event of an epidemic outbreak, the fraction of infected individuals grows until reaching a certain maximum, after which it decays to zero, if given sufficient time. When there is no epidemic outbreak, however, the number of infected (and therefore removed) individuals is relatively sufficiently low, and the number of susceptible individuals is relatively sufficiently high, that their plots versus time reveal no change after a very short time, and the respective graphs appear as horizontal lines. This may happen even with no immunization, for some network realizations or owing to the low degree or cluster size of the first infected individual. The fraction of such occurrences grows higher with immunization, and higher still as the immunized fraction of the population grows, in each of the three immunization methods that were tested. In fact, a very similar endemic fraction can be detected in randomly immunizing 10% of the population and in acquaintance immunization of 1%.

In Figure 15.4, it can be seen that for a low immunization fraction, the dynamics of random immunization is very similar to the regular SIR model. Targeted immunization is already highly effective – not only because in most of its realizations an endemic state does not occur, but also because when the epidemic does spread, a significantly smaller fraction of individuals is infected. Acquaintance immunization shows a higher fraction of endemic states than those targeted, but its fraction of infected individuals is still lower than that obtained with the random immunization strategy.

15.5 Conclusion

In conclusion, we have described several immunization methods. The random immunization method is inefficient for networks with broad degree distributions. Targeted immunization is efficient, but requires knowledge of the node degrees. The acquaintance immunization strategy is an efficient strategy for immunization, requiring no knowledge of the nodes' degrees or any other global information. This strategy is efficient for networks of any broad-degree distribution and allows for a low threshold of immunization, even where random immunization requires the entire population to be immunized. We discussed analytical results for the critical immunization fraction in both a static model and the kinetic SIR model.

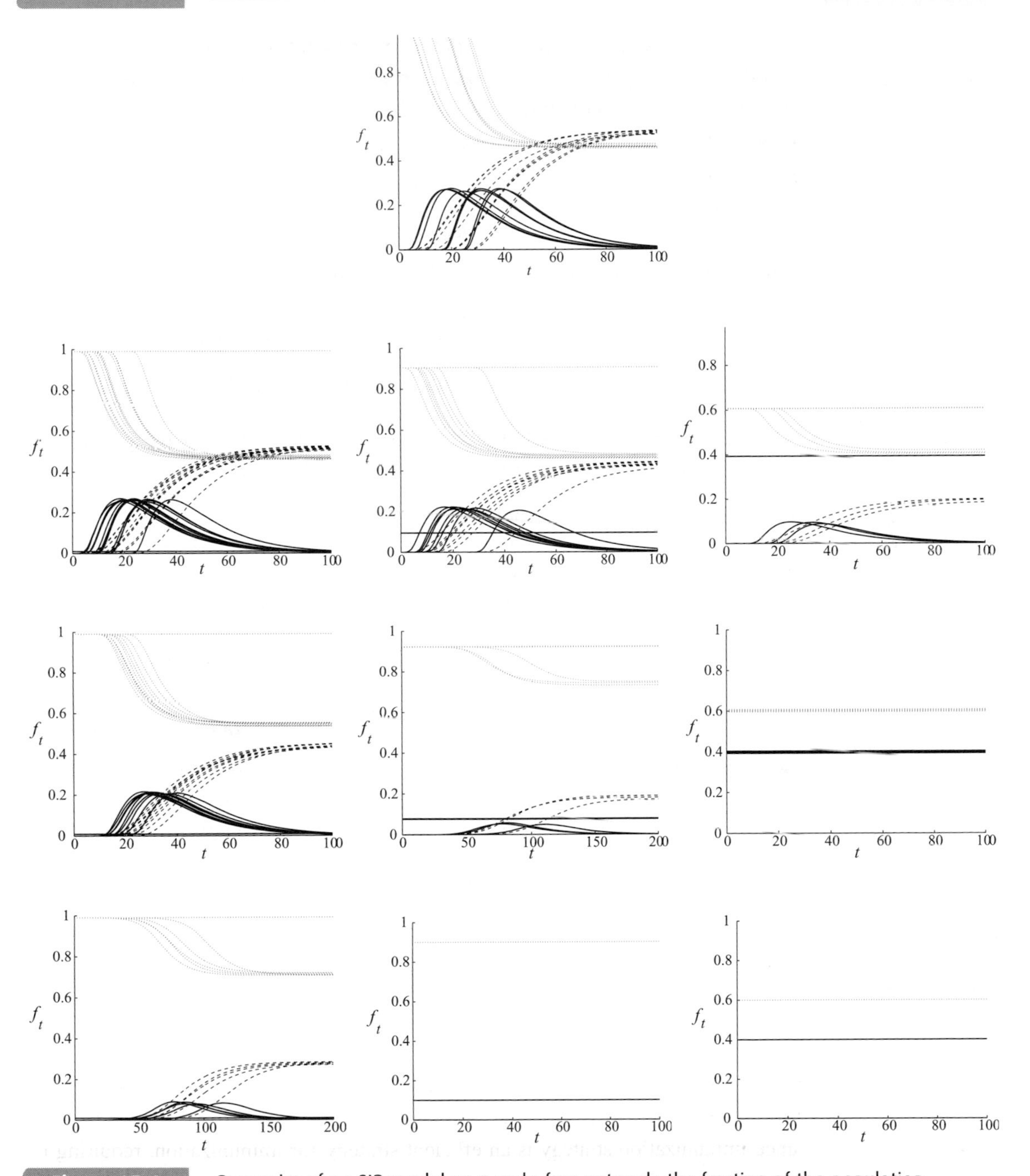

Figure 15.4 Dynamics of an SIR model on a scale-free network: the fraction of the population occupying each state, as a function of time. The graphs show the fractions of susceptible (dotted light gray line), infected (solid black line), and removed (dashed darker gray) individuals, for the regular SIR model (top), random immunization (second row), acquaintance immunization (third row), and targeted immunization (bottom), where p is 1% (left), 10% (middle column), and 40% (right). The immunized fraction is denoted by a black straight line. In all cases, the size of the network is 10^5, with $\gamma = 2.5$ and $m = 1$. The parameters of the SIR model are $r = 0.1$ and $\tau = 10$. Each graph contains 100 different realizations of the scale-free network. After [MKC+04].

As a final remark, we note that the presented approaches may be relevant to other networks, such as ecological predator–prey networks [CGA02, SM01], metabolic networks [JTA$^+$00], networks of cellular proteins [JMBO01], and terrorist networks.

Exercises

15.1 What is the immunization threshold for a random k-regular graph?

15.2 What is the immunization threshold for a k-ring?

15.3 What is the immunization threshold for a small-world network with an underlying k-ring structure (see Section 4.2)?

15.4 What is the threshold for random immunization of a two-dimensional grid (use results from two-dimensional percolation theory, e.g., [BH96, SA94])? Can you find a good targeted immunization strategy? What is the threshold for targeted immunization?

15.5 What is the effect of adding shortcuts to the two-dimensional grid on the targeted immunization threshold? How would you expect long-range flights to affect epidemic spreading and immunization considerations in human populations?

16 Thermodynamic models on networks

16.1 Introduction

In this chapter we consider the behavior of some thermodynamic models on networks. Thermodynamic models attempt to describe the macroscopic (large-scale) properties of systems by their microscopic (small-scale) behavior. The microscopic behavior stems from the interactions between atoms or molecules in the material. In the nineteenth century a new branch in physics, called "statistical physics" or "statistical mechanics" appeared (mainly owing to the works of Gibbs, Maxwell and Boltzmann as well as later works by Einstein), which tried to explain the observed properties of materials (phase transitions, magnetism, heat capacity, pressure and temperature) by the behavior of the material at the microscopic level. These studies showed that the phenomena of heat and the material properties were not basic phenomena as thought before, but rather manifestations of the small-scale interactions in the material. The microscopic level interaction coupled with statistical analysis to describe the properties of large-scale samples could explain all these phenomena and account for the large differences between materials. A well-known example is the Ising model (see, e.g., [Bax82]), dating back to the early twentieth century, which tried to explain the phenomena of ferromagnetism (the formation of an internal macroscopic magnetic moment in materials) and the transition between paramagnetism and ferromagnetism.

Many of the models studied assume discrete microscopic units of the materials representing atoms or molecules interacting with each other. The interactions are usually assumed to be local, and therefore the units were treated as nodes in some interaction network, where interaction is limited to nearest neighbors. Usually in the physics literature, thermodynamic models are studied on lattices, representing a solid state crystal of some material. Since studying such models in finite dimension tends to be very difficult owing to induced correlations, many of the models studied have been approximated by studying their mean-field behavior, or their behavior on high dimensional grids. In this chapter we will study the behavior of thermodynamic models on complex networks. This behavior will give the mean-field behavior for networks with narrow degree distribution, and will also give new universality classes and results for more heterogeneous networks, such as scale-free networks. These results may also be relevant to new kinds of materials [ZKM$^+$03] that may be produced in the near future and possess a broad degree distribution (see also [JKBH08]).

16.2 The Ising model in complex networks

The simplest thermodynamic model is the Ising model, proposed originally in the 1920s. The model studies the phase transition with the change in temperature of a paramagnet (having no internal magnetic moment, but responding to an external magnetic field) to a ferromagnet (having an internal magnetic moment). The material is modeled by a network, in which each node possesses an internal magnetic moment in a single direction (a classical simplification of a spin). In this model a node i has spin s_i, which can accept only two values, $+1$ and -1, representing magnetization in the positive and negative directions of some axis, respectively. The internal magnetic moment $\langle M \rangle = \frac{1}{N}\sum_i s_i$ is the order parameter, attaining the value $\langle M \rangle = 0$ for the paramagnetic phase and a value $0 < |\langle M \rangle| \leq 1$ for the ordered, ferromagnetic phase.

It was well known that in one dimension the Ising model shows no "physical" phase transition, i.e., only at zero temperature does a significant magnetic moment appear. The mean-field solution (the fully connected graph), on the other hand, gives a phase transition at a finite temperature. In the two-dimensional case and in higher dimensions it has been shown that a phase transition occurs at a finite temperature [Ons44].

We now turn to investigate the behavior of the Ising model in complex networks. The Hamiltonian of the Ising model is given by

$$\mathcal{H} = -\sum_{i \leftrightarrow j} J s_i s_j - \sum_i H s_i \,, \tag{16.1}$$

where the first summation is over all connected node pairs i and j. The interaction, J, and the magnetic field, H, are assumed to be uniform.

The partition function of the Ising model on a network can be written as

$$Z = \sum_{\{s_i = \pm 1\}} \exp(-\beta \mathcal{H}) \,, \tag{16.2}$$

where the sum ranges over all possible configurations for all spins $s_i = \pm 1$ and $\beta = 1/kT$, where T is the temperature and k is Boltzmann's constant. For simplicity, we will work in units in which $k = 1$ and $\beta = T^{-1}$. Using the fact that the network is locally tree like (following [DGM02b]) one can obtain exact results for the magnetization. Consider a tree rooted at node 0 (which can be any arbitrary node), where node 0 has spin s_0. Node 0 has k_0 neighbors, with spins denoted by $s_{1,i}$ where

$i = 1, 2, \ldots, k_0$ (the subscript 1 indicates the fact that these nodes are first neighbors of node 0). The contribution of the subtree rooted at spin $1, i$ (neighboring node 0) to the partition function is given by

$$z_{1,i}(s_0) = \sum_{s_l=\pm 1} \exp\left[\beta\left(\sum_{l\leftrightarrow m} s_l s_m + s_0 s_{1,i} + H \sum_l s_l\right)\right] . \quad (16.3)$$

The indices l and m run over all spins in the subtree rooted by node $1, i$ (including node $1, i$). Note that the partition function, Z, is given by

$$Z = \prod_{i=1}^{k_0} z_{1,i}(1) + \prod_{i=1}^{k_0} z_{1,i}(-1) . \quad (16.4)$$

Denoting $x_{1,i} = z_{1,i}(+1)/z_{1,i}(-1)$, the magnetic moment of node 0 is given by

$$M_0 = \frac{\mathrm{d}\ln Z}{\mathrm{d}\beta} = \frac{\mathrm{e}^{2\beta H} - \prod_{i=1}^{k_0} x_{1,i}}{\mathrm{e}^{2\beta H} + \prod_{i=1}^{k_0} x_{1,i}} . \quad (16.5)$$

The quantities $x_{1,i}$ describe the effect of the spins neighboring node 0 on its magnetic moment. Similarly, these quantities can be derived recursively by the following relation,

$$x_{n,j} = y\left(\prod_{l=1}^{k_{n,j}-1} x_{n+1,l}\right) , \quad (16.6)$$

where the product goes over all outgoing links and the function

$$y(x) = \frac{\mathrm{e}^{\beta(-1+H)} + x\mathrm{e}^{\beta(1-H)}}{\mathrm{e}^{\beta(1+H)} + x\mathrm{e}^{\beta(-1-H)}} , \quad (16.7)$$

is introduced. If node $n + 1, l$ is of degree one then $x_{n+1,l} = 1$. One can introduce an average effective magnetic field, $h_{n,l}$, defined by $x_{n,l} = \exp(-h_{n,l})$. Then, Eqs. (16.5) and (16.6) take the form

$$M_0 = \frac{\mathrm{e}^{2\beta H} - \exp(-\sum_{l=1}^{k_0} h_{1,l})}{\mathrm{e}^{2\beta H} + \exp(-\sum_{l=1}^{k_0} h_{1,l})} , \quad (16.8)$$

and

$$h_{n,j} = -\ln\left\{y\left[\exp\left(-\sum_{l=1}^{k_{n,j}-1} h_{n+1,l}\right)\right]\right\} . \quad (16.9)$$

When no external field is applied, $H = 0$, all effective fields $h_{n,l} = 0$ in the paramagnetic phase. However, in the ferromagnetic phase a spontaneous field appears and $h_{n,l} \neq 0$.

The equations derived above are valid for every tree-like structure. When discussing random networks the distribution of the node degrees should be incorporated.

Furthermore, the effective field far from a spin is assumed to be constant, as the network is well mixed, and therefore the influence of other spins is averaged in the same way at a large enough distance regardless of the initial node chosen. Introducing

$$\Psi_n(h) = \left\langle \delta(h - h_{n,l}) \right\rangle , \tag{16.10}$$

and its Laplace transform

$$\tilde{\Psi}_n(h) = \int_0^\infty dh e^{-sh} \Psi_n(h) , \tag{16.11}$$

one obtains

$$\delta\left(\sum_{j=1}^{k} h_{n,j} - h\right) = \int_{-\mathrm{i}\infty}^{\mathrm{i}\infty} \frac{\mathrm{d}s}{2\pi \mathrm{i}} \mathrm{e}^{sh} \tilde{\Psi}_n^k(s) , \tag{16.12}$$

due to the convolution property of the Laplace transform.

The average magnetic moment is

$$\langle M \rangle = \sum_k P(k) = \int_0^\infty \mathrm{d}h \frac{\mathrm{e}^{2\beta H} - \mathrm{e}^{-h}}{\mathrm{e}^{2\beta H} + \mathrm{e}^{-h}} \int_{-\mathrm{i}\infty}^{\mathrm{i}\infty} \frac{\mathrm{d}s}{2\pi \mathrm{i}} \mathrm{e}^{sh} \tilde{\Psi}_1^k(s) . \tag{16.13}$$

Using Eq. (16.9) one obtains the recurrence relation

$$\tilde{\Psi}_n(s) = \sum_k \frac{kP(k)}{\langle k \rangle} \int_0^\infty \mathrm{d}h y^s(\mathrm{e}^{-h}) \int_{-\mathrm{i}\infty}^{\mathrm{i}\infty} \frac{\mathrm{d}s'}{2\pi \mathrm{i}} \mathrm{e}^{sh} \tilde{\Psi}_{n+1}^k(s') . \tag{16.14}$$

At a dead end $\tilde{\Psi}(s) = 1$. Starting at a dead end and increasing n towards $n \to \infty$, the recurrence, Eq. (16.14), converges to a function $\tilde{\Psi}(s)$. Solving Eq. (16.14) for $\tilde{\Psi}_n(s) = \tilde{\Psi}_{n+1}(s) = \tilde{\Psi}(s)$ and substituting the result into Eq. (16.13) leads to the value of $\langle M \rangle$ for the given temperature and external field.

16.2.1 The critical temperature

For $H = 0$ in the paramagnetic phase, observed at high temperatures, $\tilde{\Psi}(s) = 1$. At a lower temperature a phase transition occurs, and another solution appears. To find the critical temperature, consider a starting function $\Psi_n(s) = \mathrm{e}^{-\delta s}$, where δ is small, such that $0 < \delta \ll 1$ and $\delta s \ll 1$. After m steps of the recurrence, Eq. (16.14), one obtains

$$\Psi_{n-m}(s) = \exp(-\delta s f^m) , \tag{16.15}$$

with

$$f = (\kappa - 1) \tanh \beta . \tag{16.16}$$

Thus, if $f = (\kappa - 1)\tanh\beta < 1$, $f^m \to 0$ and therefore the solution $\tilde{\Psi}(s) = 1$ is stable, whereas if $f > 1$, this solution is unstable and a different, non-trivial, solution appears. Hence, the critical temperature, T_c satisfies the equation

$$\tanh(1/T_c) = (\kappa - 1)^{-1} . \tag{16.17}$$

Notice that, in the case $\kappa \to \infty$ the critical temperature diverges, indicating that the system is in the ordered, ferromagnetic, phase at any finite temperature. This is very similar to the results regarding the critical percolation threshold obtained in Chapter 10.

At zero temperature, the solution of Eq. (16.14) is $\tilde{P}si(s) = t_s + (1 - t_s)\delta_{s,0}$, where t_s is the the smallest root of $\sum_k kP(k)t_s^{k-1}$. The magnetization is then

$$\langle M \rangle = 1 - \sum_k P(k)t_s^k , \tag{16.18}$$

which is exactly the size of the giant component, Eq. (9.8). This is expected, as at zero temperature all the spins in any given component freeze in the same direction. However, spins in different components have no interaction and therefore their respective directions are random. Thus, the only macroscopic contribution is from the giant component.

16.2.2 The critical exponents

To study the critical exponents, one can use the effective medium approach. In this approach, it is assumed that each spin's neighbors behave as a random sample of all the spins in the network. This leads to the approximation

$$\sum_{l=1}^{k} h_{n,l} \approx kh + \mathcal{O}(k^{1/2}) , \tag{16.19}$$

with $h = \langle h_{n,l} \rangle$ being the average effective field. This leads to

$$\langle M \rangle = \sum_k P(k) \frac{e^{2\beta H} - e^{-kh}}{e^{2\beta H} + e^{-kh}} , \tag{16.20}$$

where h is obtained by substituting the approximation, Eq. (16.19), into Eq. (16.9) to obtain

$$h = -\sum_k \frac{kP(k)}{\langle k \rangle} \ln\left(y\left(e^{-(k-1)h}\right)\right) \equiv G(h) . \tag{16.21}$$

Expanding $G(h)$ in the case $\gamma > 5$ one obtains $G(h) = g_1 h + g_3 h^3 + \cdots$, with g_1, g_3 being some constants that depend on the temperature. Substituting this into

Eq. (16.21) leads to an equation of the form $h = ch^3$, where c is a constant depending on the temperature. Expanding near the critical point, where h and $T - T_c$ are small leads to

$$h \approx \left(\frac{12 \langle k(k-1) \rangle^2}{\langle k \rangle \langle k(k-1)^3 \rangle} T_c \tau \right)^{1/2} \equiv a\tau^{1/2} , \tag{16.22}$$

where $\tau = 1 - T/T_c$ is the reduced temperature. Equation (16.20) then results in

$$\langle M \rangle \approx \langle k \rangle \frac{a\tau^{1/2}}{2} . \tag{16.23}$$

The behavior of the magnetization near the critical temperature is therefore, $\langle M \rangle = 0$ for $T > T_c$, and $\langle M \rangle \propto \tau^{1/2}$ for $T < T_c$. This is the same result as obtained in the regular mean-field Ising model in infinite dimension.

The magnetic susceptibility is obtained by expanding $\langle M \rangle$ with a magnetic field and then taking the derivative of $\chi = \mathrm{d}\langle M \rangle / \mathrm{d}H$ at $H = 0$. For $T < T_c$ this results in

$$\chi(H=0) \approx \frac{\langle k \rangle^3}{2 \langle k^2 \rangle \langle k(k-2) \rangle} \tau^{-1} . \tag{16.24}$$

At $T > T_c$ the behavior $\chi \propto \tau^{-1}$ is conserved, but the prefactor is different.

The specific heat is defined as $C = \mathrm{d}\langle E \rangle / \mathrm{d}T$, where $\langle E \rangle$ is the average energy per spin

$$\langle E \rangle = \frac{1}{N} \left\langle -J \sum_{\langle ij \rangle} S_i S_j \right\rangle_T , \tag{16.25}$$

resulting in

$$\langle E \rangle = -\frac{\langle k \rangle \coth(2/T)}{2} + \frac{1}{2\sinh(2/T)} \sum_k k P(k) \frac{\mathrm{e}^{-h} + \mathrm{e}^{-(k-1)h}}{1 + \mathrm{e}^{-kh}} . \tag{16.26}$$

Using $h \approx a\tau^{1/2}$ one finds that the specific heat presents a jump at $T = T_c$ with

$$\Delta C = \frac{\langle k(k-2) \rangle \langle k^2 \rangle a^2}{8 \langle k \rangle} . \tag{16.27}$$

Notice that when $\langle k^4 \rangle \to \infty$, $a \to 0$ and the size of the jump approaches zero.

For scale-free networks with $3 < \gamma < 5$ the expansion of $G(h)$ yields $G(h) \approx g_1 h + g_3 h^{\gamma - 2}$. Similar to above, one obtains

$$\langle M \rangle \propto \tau^{1/(\gamma-3)} , \tag{16.28}$$

$$\delta C \propto \tau^{(5-\gamma)/(\gamma-3)} , \tag{16.29}$$

$$\chi \propto \tau^{-1} , \tag{16.30}$$

where δC is the change in specific heat with the temperature. For $T > T_c$, $\delta C = 0$.

Table 16.1 Summary of the behavior of the critical temperature

The critical behavior of the magnetization M, the specific heat δC, and the susceptibility χ in the Ising model on networks with a degree distribution $P(k) \sim k^{-\gamma}$ for various values of exponent γ. $\tau \equiv 1 - T/T_c$. The right column represents the exact critical temperature, T_c, in the case $\langle k^2 \rangle < \infty$ and the dependence of T_c on the total number N of vertices in the network. All values are up to a constant scale factor. After [DGM02b].

γ	M	$\delta C(T < T_c)$	χ	T_c
$\gamma > 5$	$\tau^{1/2}$	jump at T_c	τ^{-1}	$2/\ln \frac{\langle k^2 \rangle}{\langle k^2 \rangle - 2\langle k \rangle}$
$\gamma = 5$	$\tau^{1/2}/(\ln \tau^{-1})^{1/2}$	$1/\ln \tau^{-1}$	τ^{-1}	$2/\ln \frac{\langle k^2 \rangle}{\langle k^2 \rangle - 2\langle k \rangle}$
$3 < \gamma < 5$	$\tau^{1/(\gamma-3)}$	$\tau^{(5-\gamma)/(\gamma-3)}$	τ^{-1}	$2/\ln \frac{\langle k^2 \rangle}{\langle k^2 \rangle - 2\langle k \rangle}$
$\gamma = 3$	$e^{-2T/\langle k \rangle}$	$T^2 e^{-4T/\langle k \rangle}$	T^{-1}	$\langle k \rangle \ln N$
$2 < \gamma < 3$	$T^{-1/(3-\gamma)}$	$T^{-(\gamma-1)/(3-\gamma)}$	T^{-1}	$\langle k \rangle N^{(3-\gamma)/(\gamma-1)}$

For $\gamma < 3$ the critical temperature diverges. Using the expansion $G(h) \sim gh^{\gamma-2}/T$ one obtains

$$\langle M \rangle \propto T^{-1/(3-\gamma)} , \tag{16.31}$$

$$\delta C \propto T^{-(\gamma-1)/(3-\gamma)} , \tag{16.32}$$

$$\chi \propto T^{-1} . \tag{16.33}$$

A summary of the results for the behavior of the critical temperature and critical exponents is presented in Table 16.1 Notice that for the separating cases, $\gamma = 3, 5$, logarithmic corrections occur.

16.3 Summary

In summary, the scale-free nature of a network has a strong influence on the thermodynamic behavior of the Ising model, in a very similar manner to the effect of the scale-free nature on the percolation behavior. For scale-free networks with $\gamma \leq 3$ the ordered phase dominates, and no phase transition occurs at any finite temperature. For $3 < \gamma \leq 5$ a phase transition occurs at a finite temperature. However, the critical exponents differ from the mean-field exponents. This is comparable to the non-universal behavior appearing in percolation, with the difference being that in percolation mean-field behavior is restored for $\gamma > 4$.

Similar behavior is observed for other thermodynamical models, see, for example, [GDM02b, DG08] for results on other models.

Exercises

16.1 Obtain the coefficients in the expansion of $G(h)$, Eq. (16.21), and derive Eq. (16.22).

16.2 Obtain the susceptibility of the Ising model for $\gamma < 5$, Eq. (16.24), and the similar expression for $T > T_c$.

16.3 Derive Eqs. (16.28)–(16.30).

16.4 Derive Eqs. (16.31)–(16.33).

16.5 Solve the Ising model on a random r-regular network (with all nodes having degree r). Obtain the expressions for T_c and the critical exponents.

17 Spectral properties, transport, diffusion and dynamics

In this chapter we discuss the spectral properties of networks, and their relation to dynamical properties such as diffusion. There are two main characteristic matrices for a graph, the adjacency matrix and the Laplacian. We discuss both of them and explain the relation to the dynamical properties. A good summary of results on the spectrum of networks can be found in [Chu97]. For a nice survey of random walks on graphs see [Lov96].

17.1 The spectrum of the adjacency matrix

As described in Chapter 1, the adjacency matrix, A, is an $N \times N$ matrix (where N is the number of nodes) whose entries are $A_{i,j} = 1$ if there is an edge between nodes i and j, and $A_{i,j} = 0$ otherwise. If the graph is not directed then the adjacency matrix is symmetric. In a directed graph it is not necessarily so.

The **spectrum** of this matrix is the collection of eigenvalues of the matrix. Since we refer to an ensemble of graphs rather than a single graph, we are interested in the distribution of eigenvalues when the entire ensemble is considered.

The study of the spectrum of random graphs is strongly related to the study of the spectrum of random matrices in general. The study of random matrix theory was initiated by Wigner [Wig55] in the 1950s. His main goal was to understand the energy levels of the nucleus. These levels are the eigenvalues of the Hamiltonian (energy) matrix. Wigner's reasoning was that in large nuclei the interactions are complicated enough that the matrix entries can be considered to be randomly selected. The only condition is that the matrix must obey the symmetries of the physical system. The study of random matrices was found useful also in the study of localization of electrons and photons in solid state physics [Hof76]. Some results regarding localization in complex networks can be found in [JKBH08, SKHB05].

The study of the spectrum of large random matrices led to the discovery of the celebrated **semicircle law** [Wig55]. This law asserts that for a large real symmetric $N \times N$ matrix with random independent entries taken from a distribution having second moment m^2 and all other moments converging, the probability density of having an eigenvalue x is

$$p(x) = \frac{1}{2\pi m^2 N}\sqrt{4m^2 N - x} \ . \tag{17.1}$$

The spectrum of ER networks with large average degree (the average degree of a random graph must be at least $\langle k \rangle > \ln N$ for the graph to be connected, i.e., to have only one component) follows the same semicircle law with $m^2 = Np(1-p)$, where p is the probability of a link being occupied. If the average degree is not too high $\langle k \rangle \ll N$, $1 - p \approx 1$ and therefore $m^2 \approx \langle k \rangle$. The spectrum of the adjacency matrix of scale-free graph models has been studied in several papers [FDBV01, FFF99, GKK01a]. The spectrum is different from the spectrum of ER graphs. It has been found that the tail of the spectrum of a scale-free graph also decays as a power law.

17.2 The Laplacian

Consider a diffusion process or a random walk on the nodes of the graph. A good survey of random walks on a graph can be found in [Lov96]. The equations for a discrete time random walk on a graph are

$$P_{t+1}(i) = \sum_j \frac{P_t(j)}{k_j} \ , \tag{17.2}$$

where $P_t(i)$ is the probability of being at node i at time t and the summation ranges over all the j neighboring i. If we mark by $\pi_t = (P_t(1), P_t(2), \ldots)^t$ the vector of probabilities of being at each node at time t, then Eq. (17.2) can be rewritten in the matrix notation $\pi_{t+1} = M\pi_t$, where

$$M = \begin{pmatrix} 0 & \frac{A_{1,2}}{k_2} & \cdots \\ \frac{A_{1,2}}{k_1} & 0 & \cdots \\ \vdots & \vdots & \ddots \end{pmatrix} , \tag{17.3}$$

where A is the adjacency matrix. The configuration at time τ, starting from a configuration π_0 at time $t = 0$ can be written as

$$\pi_\tau = M^\tau \pi_0 \ . \tag{17.4}$$

If the graph is fully connected (that is, has no isolated components) then Eq. (17.4) has only one positive fixed point,

$$\pi = \begin{pmatrix} \frac{k_1}{N\langle k \rangle} \\ \frac{k_2}{N\langle k \rangle} \\ \vdots \end{pmatrix} . \tag{17.5}$$

This solution corresponds to the eigenvector of M with eigenvalue 1. If more than one component exists, then each random walk is restricted to its own component, and the space of fixed points contains all combinations of the individual fixed points for each component.

The problem with the matrix M is that it is not symmetric. This implies that its left and right eigenvectors are not identical. To simplify the treatment the following matrix can be constructed,

$$N = D^{1/2} A D^{-1/2} = \begin{pmatrix} 0 & \frac{A_{1,2}}{\sqrt{k_1 k_2}} & \cdots \\ \frac{A_{1,2}}{\sqrt{k_1 k_2}} & 0 & \cdots \\ \vdots & \vdots & \ddots \end{pmatrix}, \tag{17.6}$$

where $D = \text{diag}(k_1, k_2, \ldots)$ is the matrix having the node degrees along the diagonal and zeros off-diagonal. The matrix N is symmetric and therefore has the same left and right eigenvectors. Notice that $M = D^{-1/2} N D^{1/2}$, and thus,

$$M^\tau = (D^{-1/2} N D^{1/2})^\tau = D^{-1/2} N^\tau D^{1/2} . \tag{17.7}$$

This allows Eq. (17.4) to be written as

$$\pi_\tau = D^{-1/2} N^\tau D^{1/2} \pi_0 . \tag{17.8}$$

This implies that the behavior of the diffusion equation is controlled by the spectrum of the matrix N.

Other matrices of interest are the **Laplacian**, L, defined by the equation

$$L = D - A = \begin{pmatrix} k_1 & -A_{1,2} & \cdots \\ -A_{1,2} & k_2 & \cdots \\ \vdots & \vdots & \ddots \end{pmatrix}, \tag{17.9}$$

and the **normalized Laplacian** (sometimes also called simply the Laplacian),

$$\mathcal{L} = I - N = \begin{pmatrix} 1 & -\frac{A_{1,2}}{\sqrt{k_1 k_2}} & \cdots \\ -\frac{A_{1,2}}{\sqrt{k_1 k_2}} & 1 & \cdots \\ \vdots & \vdots & \ddots \end{pmatrix}. \tag{17.10}$$

The Laplacian reduces to the regular, continuous Laplacian (up to a constant factor), $L = \nabla^2$, when considering a d-dimensional lattice and taking the limit of very small lattice spacing.

17.3 The spectral gap and diffusion on graphs

As described in Section 17.2, the properties of diffusion on networks are determined by the spectrum of the matrix N, defined in Eq. (17.6). Since this is a symmetric matrix with all non-negative elements, the Perron–Frobenius theorem [BP94] guarantees that it has one eigenvector with all non-negative elements, and this vector has the largest eigenvalue.[1] The largest eigenvalue is therefore 1, similarly to the largest eigenvalue for M as described above, and the appropriate eigenvector is,

$$\mathbf{v} = \begin{pmatrix} \frac{\sqrt{k_1}}{\sqrt{N \langle k \rangle}} \\ \frac{\sqrt{k_2}}{\sqrt{N \langle k \rangle}} \\ \vdots \end{pmatrix} . \tag{17.11}$$

This is also an eigenvector of the normalized Laplacian with eigenvalue $\lambda_v = 0$. The eigenvalues of the normalized Laplacian, λ'_i obey the relation $\lambda'_i = 1 - \lambda_i$, where λ_i are the eigenvalues of N, and therefore, for the normalized Laplacian, $0 \leq \lambda'_i \leq 2$.

For the non-normalized Laplacian the smallest eigenvalue is given by the vector

$$\mathbf{1} = \begin{pmatrix} 1 \\ 1 \\ \vdots \end{pmatrix} . \tag{17.12}$$

As L is symmetric, all other eigenvectors are perpendicular to $\mathbf{1}$. Each eigenvector can be considered as a function on the nodes, assigning to each node i the value of the appropriate entry in the vector, $f(i)$. If $f(i)$ is an eigenfunction of L then $g = D^{1/2} f$ is an eigenfunction of $\mathcal{L}$. It then follows that

$$\frac{g \cdot \mathcal{L} g}{g \cdot g} = \frac{f \cdot L f}{(D^{1/2} f) \cdot (D^{1/2} f)} = \frac{\sum_{i \leftrightarrow j} (f(i) - f(j))^2}{\sum_i f(i)^2 k_i} , \tag{17.13}$$

where the sum on the numerator goes over all pairs of adjacent nodes, i and j.

[1] This can easily be seen in this symmetric case. Assume that v is a normalized eigenvector with some negative elements and respective eigenvalue λ_v, and u is the vector obtained by the absolute values of the elements of v. Then, by positivity, every element of Nu is at least as large in absolute value as the respective element of Nv. Therefore, $|u^t N u| \leq |v^t N v|$, and thus there must be at least one eigenvector with eigenvalue larger than $|\lambda_v|$.

17.3.1 The spectral gap

From Eq. (17.4) it can be seen that the behavior of random walks and diffusion processes is determined by the behavior of powers of the matrix M or, alternatively, of the Laplacian. A simple method of determining the behavior of powers of a matrix is to consider its eigenvalues. Expressing the vector π_0 by the eigenvectors of the normalized Laplacian,

$$\pi_0 = \sum_i \alpha_i v_i \,, \tag{17.14}$$

where v_i are the eigenvectors of the matrix M and α_i are the projections of π_0 on the different eigenvectors, leads to

$$\pi_t = M^t \sum_i \alpha_i v_i = \sum_i \alpha_i \lambda_i^t v_i \,. \tag{17.15}$$

Since for all eigenvalues $-1 \le \lambda_i \le 1$ it follows that the absolute value of all summands decays exponentially, except for those corresponding to eigenvalues ± 1. If the graph is connected, only one eigenvalue can be equal to one. The eigenvalues are usually ordered in decreasing size, and therefore this eigenvalue is denoted by λ_1, and

$$1 = \lambda_1 \ge \lambda_2 \ge \cdots \ge \lambda_N \ge -1 \,. \tag{17.16}$$

An eigenvalue of -1 indicates that the diffusion alternates between two sets of nodes. This happens only when the graph is bipartite, i.e., the graph can be divided into two sets of nodes, where no links exist between any two nodes in the same set. One can consider a diffusion process in which the walker has a probability $1/2$ of staying in its current location at every step. This is also equivalent to a graph in which we add k self loops to every node of degree k. In this case the eigenvalues of the diffusion are all positive (or at least non-negative),

$$1 = \lambda_1 \ge \lambda_2 \ge \cdots \ge \lambda_N \ge 0 \,. \tag{17.17}$$

From Eq. (17.15) it follows that except for the eigenvector corresponding to eigenvalue 1 (and possibly the eigenvector corresponding to eigenvalue -1, for bipartite graphs) all other eigenstates decay exponentially in time.

Following Eq. (17.13), the second smallest eigenvalue of $\mathcal{L}$ is given by

$$\lambda_1' = \min_{f \perp D1} \frac{\sum_{i \leftrightarrow j} (f(i) - f(j))^2}{\sum_i f(i)^2 k_i} \,. \tag{17.18}$$

17.3.2 Expander graphs

An interesting and important class of graphs is the class of expander graphs. Expander graphs are graphs in which the size of a cut needed to separate a set of nodes from the rest of the graph is proportional to the size of the set. That is, a large group of nodes cannot be separated from the rest of the graph by cutting a small number of links. This is similar to the property of infinite-dimensional networks discussed, for example, in Chapter 7. However, the expansion property is stronger, as it applies to every set of nodes in the network, and not only on average. For instance, a network with finite components cannot be an expander, as formally they can be separated from the rest of the network by cutting zero links. However, it can still be infinite dimensional, as its giant component can still show good expansion properties (at least on average).

Formally, consider a set of nodes, S, and mark by $\overline{S}$ the complementary set, i.e., the set of all nodes not in S. We denote by vol$S$ the volume of S, defined as the total number of links of all nodes in S,

$$\mathrm{vol}S \equiv \sum_{i\in S} k_i \,. \tag{17.19}$$

The Cheeger constant of a set of nodes[2] is defined as [Chu97]

$$\Phi(S) = \frac{|E(S,\overline{S})|}{\min(\mathrm{vol}S, \mathrm{vol}\overline{S})}\,, \tag{17.20}$$

where $|E(S,\overline{S})|$ denotes the number of edges connecting between nodes in S and nodes in $\overline{S}$. The **conductance** of a graph is defined by

$$\Phi = \min_S \frac{|E(S,\overline{S})|}{\min(\mathrm{vol}S, \mathrm{vol}\overline{S})}\,. \tag{17.21}$$

If Φ is a constant independent of N, then the graph is said to be an expander graph. Expander graphs play an important role in mathematics and computer science, as they have several important properties. Among others, they have considerably good resilience to both random and targeted attacks on the links, as a large number of links needs to be removed in order to disconnect a large chunk of the network. In addition, they have a relatively large spectral gap, and therefore offer good mixing. Thus, processes on expander graphs quickly approach the equilibrium state. Expander graphs also allow good routing strategies [GMS03, Vaz04] and have numerous other applications from circuit design to error correcting codes.

[2] The Cheeger constant in graphs is analogous to the Cheeger constant in manifolds.

To find the relation between the conductance and the spectral gap consider the following [Chu97]. Suppose the minimum of Eq. (17.20) is obtained for some set S_1, and that the number of edges between S_1 and $\overline{S_1}$ is c. Define the following function on the nodes

$$f(i) = \begin{cases} \frac{1}{\mathrm{vol} S_1} & i \in S_1 \\ -\frac{1}{\mathrm{vol}\overline{S_1}} & i \notin S_1 \, . \end{cases} \tag{17.22}$$

Obviously, $f \perp D\mathbf{1}$. Equation (17.18) then implies that

$$\lambda_1' \le c(1/\mathrm{vol} S_1 + 1/\mathrm{vol}\overline{S_1}) \le \frac{2c}{\min(\mathrm{vol} S_1, \mathrm{vol}\overline{S_1})} = 2\Phi \, . \tag{17.23}$$

This gives an upper bound on λ_1'.

To obtain a lower bound consider the function f which gives the eigenvalue λ_1'.[3] Suppose that we sort the nodes according to increasing value of f, such that for all i, $f(i) \le f(i+1)$. Assume that $\sum_{f(i)<0} k_i \ge \sum_{f(j)\ge 0} k_j$.[4] For each i consider the cut between S_i, the set of nodes with indices above i and $\overline{S_i}$, the set of nodes with indices below i including i. Let S be the set of nodes with $f(i) < 0$, and define

$$g(i) = \begin{cases} f(i) & f(i) > 0 \\ 0 & f(i) \le 0 \, . \end{cases} \tag{17.24}$$

Denote by r the index for which $\mathrm{vol} S_{k-1} \le N \langle k \rangle /2$ and $\mathrm{vol}\bar{S}_k \le N \langle k \rangle /2$. Denote $\pi(i) = k_i/(N\langle k \rangle)$.

For any function $h(i)$ satisfying $\mathrm{vol}\{i | h(i) > 0\} \le N \langle k \rangle /2$ and $\mathrm{vol}\{i | h(i) < 0\} \le N \langle k \rangle /2$ we have

$$\begin{aligned} \sum_{i \leftrightarrow j} |h(i) - h(j)| &= \sum_{i=1}^{N-1} |E(S_i, \bar{S}_i)|(h(i+1) - h(i)) \\ &\ge \langle k \rangle \, \Phi \sum_{i=1}^{N-1} (h(i+1) - h(i)) \frac{\mathrm{vol} S_i \mathrm{vol}\bar{S}_i}{N} \, . \end{aligned} \tag{17.25}$$

[3] In the following we closely follow [Lov96, Sin93].

[4] If this inequality does not hold consider the function $-f$, for which it will hold, and which will give the same eigenvalue.

Let t be the index such that $\mathrm{vol}S_t \leq N/2$ and $\mathrm{vol}S_{t+1} \geq N/2$. We have

$$
\begin{aligned}
\sum_{i\leftrightarrow j} |h(i) - h(j)| \\
&\geq \frac{N\langle k\rangle \Phi}{2} \sum_{i=1}^{t} (h(i+1) - h(i))\mathrm{vol}S_i \\
&\quad + \frac{N\langle k\rangle \Phi}{2} \sum_{i=1}^{t} (h(i+1) - h(i))\mathrm{vol}\bar{S}_i \\
&= \frac{N\langle k\rangle \Phi}{2} \sum_i \pi(i)|h(i)| = \frac{N\langle k\rangle \Phi}{2}\,. \qquad (17.26)
\end{aligned}
$$

Now,

$$
\begin{aligned}
\sum_i \pi(i)g(i)^2 &\geq \frac{\sum_i \pi(i)(f(i) - f(k))^2}{2} \\
&= \frac{\sum_i \pi(i)f(i)^2}{2} - f(k)\sum_i \pi(i)f(i) + \frac{f(k)^2}{2} \\
&= \frac{1}{N\langle k\rangle} + \frac{f(k)^2}{2} \geq \frac{1}{N\langle k\rangle}\,. \qquad (17.27)
\end{aligned}
$$

Using Eq. (17.26) with the numbers $h(i) = g(i)^2/\sum_i(\pi(i)g(i)^2)$ one obtains

$$
\sum_{i\leftrightarrow j} |g(i)^2 - g(j)^2| \geq N\langle k\rangle \sum_i \pi(i)g(i)^2\,. \qquad (17.28)
$$

On the other hand

$$
\begin{aligned}
\sum_{i\leftrightarrow j} |g(i)^2 - g(j)^2| \\
&\leq \left(\sum_{i\leftrightarrow j}(g(i) - g(j))^2\right)^{1/2} \left(\sum_{i\leftrightarrow j}(g(i) + g(j))^2\right)^{1/2}. \qquad (17.29)
\end{aligned}
$$

The second factor can be bounded using

$$
\sum_{i\leftrightarrow j}(g(i) + g(j))^2 \leq 2\sum_{i\leftrightarrow j}(g(i)^2 + g(j)^2) = 4N\langle k\rangle \sum_i \pi(i)g(i)^2\,. \qquad (17.30)
$$

Combining Eqs. (17.29) and (17.30), one finally obtains

$$\begin{aligned}\sum_{i\leftrightarrow j}(f(i)-f(j))^2 &\geq \sum_{i\leftrightarrow j}(g(i)-g(j))^2 \geq \frac{\left(\sum_{i\leftrightarrow j}|g(i)^2-g(j)^2|\right)^2}{\sum_{i\leftrightarrow j}(g(i)+g(j))^2}\\ &\geq (\Phi N\langle k\rangle)^2\frac{\left(\sum_i \pi(i)g(i)^2\right)^2}{4N\langle k\rangle\sum_i \pi(i)g(i)^2}\\ &= \frac{\Phi^2 N\langle k\rangle}{4}\sum_i \pi(i)g(i)^2 \geq \frac{\Phi^2}{8}\,. \end{aligned} \tag{17.31}$$

Thus, we have

$$2\Phi \geq \lambda_1' \geq \frac{\Phi^2}{8}\,. \tag{17.32}$$

This inequality is the discrete version of the Cheeger inequality. From this inequality it follows that if a graph has a finite spectral gap, then it also has finite conductance and vice versa. As we have seen, a finite gap also guarantees fast mixing of a random walk. Therefore, expander graphs present many desirable properties which follow from their spectral gap.

It should be mentioned that there are other definitions for expansion, which depend on other, similar properties of the graph. For instance, the vertex (node) expansion of a graph can be considered, which measures the ratio between the number of nodes in a set and the number of nodes, rather than links, connected to this set.[5] Usually, graphs having good expansion properties using one definition also have good expansion properties based on the other definitions. This is mainly true if the degree distribution is not very broad, as then each node is only connected to a bounded number of links and therefore, expansion in links guarantees expansion in nodes and vice versa. See [SCL$^+$07] for some discussion of vertex expansion in scale-free networks and its applications to routing.

17.3.3 Expansion of random networks

In general, random graphs tend to have good expansion properties. This follows from the randomness in the connections, which leads to an almost uniform spread of the links between groups of nodes, and thus good expansion properties are obtained for most cuts in the network. However, for this property to hold, the number of edges should be large enough, so that the results concentrate around the mean. In

[5] In fact, the expansion is the inverse of this ratio: the number of neighboring nodes divided by the set size.

the following we study the expansion properties of a generalized random graph. See [GMS03] for a more complete discussion.

Consider a generalized random graph with a minimum degree m, having N nodes. The total volume of all nodes in the graph is $V = N\langle k\rangle$. Suppose now that we choose a set of nodes having s nodes and volume $\upsilon \leq V/2$. Clearly, $\upsilon \geq ms$. Now, for the graph to have expansion Φ, each such set must have at least $\Phi\upsilon$ links going to nodes outside this set. The probability this event will not occur, i.e., that the set will not have expansion Φ is

$$P_{\text{bad}}(\upsilon) = \sum_{n=0}^{\Phi\upsilon-1} P(\upsilon, n) , \tag{17.33}$$

where $P(\upsilon, n)$ is the probability that a set having volume υ will have exactly n links outgoing from the set. Since the graph is built using the configuration model (see Section 4.3), this probability corresponds to the probability of having n open links connecting to n open links in the complement set, and the rest connecting internally. Thus,

$$P(\upsilon, n) = \binom{\upsilon}{n}\binom{V}{n}\frac{f(2n)f(\upsilon - n)f(V - \upsilon - n)}{f(V)} , \tag{17.34}$$

where

$$f(x) = \frac{x!}{2^{x/2}\cdot(x/2)!} \tag{17.35}$$

is the number of *perfect matchings* on x nodes, i.e., the number of ways in which x objects can be divided into pairs. Since $P(\upsilon, n)$ increases with n, it follows that

$$P_{\text{bad}}(\upsilon) \leq \Phi\upsilon\binom{\upsilon}{\Phi\upsilon}\binom{V}{\Phi\upsilon}\frac{f(2\Phi\upsilon)f((1-\Phi)\upsilon)f(V-(1+\Phi)\upsilon)}{f(V)} . \tag{17.36}$$

Using Eq. (17.36) one can bound the probability that a bad set will exist in the network. This probability is

$$P_{\text{bad}} = \sum_{\upsilon=m}^{V/2} N_\upsilon P(\upsilon) , \tag{17.37}$$

where N_υ is the number of sets of volume υ, which is at most

$$N_\upsilon \leq \binom{V/m}{\upsilon/m} . \tag{17.38}$$

Using Stirling expansion of the binomial expression, one finds that choosing $\Phi = \mathcal{O}(1/\ln N)$ the probability $P_{\text{bad}} \to 0$, given that $m \geq 3$. For $m = 1$ or $m = 2$ the graph is not an expander, as expected, as nodes can form relatively long chains only connected to the rest of the graph through one or two links. Therefore, it is almost certain that every random graph with $m \geq 3$ has good expansion properties, and thus it is expected to present rapid mixing.

17.4 Traffic and self-similarity

The behavior of traffic in networks has been a subject of study for over a decade [PW00]. The study of network traffic concentrates on the volume of data flowing through the network, or through specific network links, as a function of time. Before the first quantitative studies, it was commonly believed that the expected randomness in network traffic would lead to a Poissonian traffic pattern. However, the first studies in this field [LTWW94] indicated that even in local area networks, which are small scale and do not present a scale-free topological structure, traffic is self-similar.

The correct modeling of network traffic is required for network performance analysis, and therefore has important consequences on the design and evaluation of communication networks. A study of a new network design, of new communication equipment, and of new protocols and algorithms for communication networks should take into account the correct model for network traffic in order to predict correctly the overall expected performance of the communication network. Thus, the incorporation of a self-similar traffic model is essential to network engineering tasks.

There are several models for the origin of self-similar traffic. For a review of self-similarity and its origins, see [PW00]. It should be noted that it is believed that in real-world networks, such as IP networks, the distribution of the sizes of communication tasks is broad, and possibly a power-law distribution. This has led to the notion of "elephants and mice" in communication networks.

17.5 Summary

The properties of diffusion and other dynamical behavior in networks strongly depend on the spectral properties of matrices related to the network, such as the Laplacian and the adjacency matrix. Unlike most properties discussed in this book, the rate of diffusion mainly depends on the lowest degree of nodes, rather than the high-degree tail of the degree distribution. Random networks with minimal degree at least 3 have good expansion properties and present fast diffusion rates. See also [GA04, SDM08] for studies of diffusion in complex networks.

Several other properties depend on the spectrum of networks, some examples include the resistance of resistor networks [LCH$^+$06], and synchronization of networks of oscillators [Boc08, PRK03, Str03].

Exercises

17.1 Find the eigenvalues of the adjacency matrix and the Laplacian spectrum for:
(a) the complete graph with N nodes,
(b) a star network containing one central node of degree $N - 1$ connected to $N - 1$ leaves of degree one,
(c) a d-dimensional grid of linear size L, containing $N = L^d$ nodes.

17.2 By writing the equations for the average time to reach a node i from a node j by a random walk, find the average time to move between two arbitrary nodes in:
(a) a ring of N nodes,
(b) a path of N nodes,
(c) a clique of N nodes.

17.3 The network presenting the worst average time to diffuse between two nodes is the lollipop graph, consisting of a clique containing $2N/3$ nodes, one of which is connected to a path of length $N/3$ nodes (see [Fei97] for a proof). Write the equations for T_i, the average time to reach the end of the path, starting at a node i of the lollipop graph, and solve them. *Hint:* Use the symmetry of the clique to reduce the number of equations.

17.4 Derive the behavior presented in Section 17.3.3.

18 Searching in networks

18.1 Introduction

One of the most important tasks in networks is searching for nodes when partial or no knowledge of their location is available. This task may be carried out in several different settings and may have many different applications. An example of searching in social networks was demonstrated in Stanley Milgram's famous experiment [Mil67], where individuals were requested to search for a path to other individuals chosen randomly from a phone book. The surprising conclusion of the experiment was that, on average, these random pairs of individuals were only separated by six nodes. However, another surprising result of the experiment was that the tested individuals were actually able to find such a short path easily in the complex network. A random walk on a network is expected to visit a large fraction of the nodes in the network before reaching the destination node. Nevertheless, the subjects of Milgram's experiment were able to locate the destination individuals with a number of steps near the actual network diameter. In this chapter we will discuss some aspects and techniques for searching nodes in complex networks.

18.2 Searching using degrees

In [ALPH01] a method is suggested for searching scale-free networks with exponent $\gamma \approx 2$. More specifically, the following was suggested. Each node contains information not only on its neighbors, but also on its next nearest neighbors, the rationale being that the average number of next nearest neighbors diverges in networks with $\gamma < 3$, whereas the average number of nearest neighbors is a constant. The search itself is performed in two stages.

(i) Starting at the source node, from every node encountered, move to its highest degree neighbor, until the largest hub in the network is reached.

(ii) Starting from the largest hub, go down the degree sequence, i.e., from every node encountered, go to its neighbor with the highest degree not yet visited. Stop when a node having information on the target node is reached.

It is argued [ALPH01] that using this technique has a high probability of success, i.e., stage (i) will always go up the degree sequence and will eventually reach the highest degree node after a small number of steps, and stage (ii) will lead to traversal of all of the nodes down the degree sequence until a node aware of the destination is located.

The reasoning behind the success of the first stage is as follows. Suppose we are currently at a node having degree k_1. The probability of each of its links leading to a node of degree k is $kP(k)/\langle k\rangle \sim k^{1-\gamma}$. Since there are k_1 such neighbors, and assuming k_1 is large, the highest degree of a neighbor, $k_{\max}$ can be approximated by:

$$\int_{k_{\max}}^{\infty} \frac{kP(k)}{\langle k\rangle} dk \approx \frac{1}{k_1} . \tag{18.1}$$

This leads to $k_{\max} \sim k_1^{1/(\gamma-2)} \gg k_1$. Therefore, stage (i) leads to a rapid ascent in the degree sequence. This is true when one starts from a high-degree node. However, in [ALPH01] it is shown numerically that this procedure works well starting from almost any node. In fact, it is easy to see that even initiating a random walk would quickly lead to some high-degree node from which stage (i) can continue.

The success of the second stage depends upon the fact that the high-degree nodes, having degree exceeding $\sqrt{N}$, form a clique with high probability (i.e., they are all connected together). When γ approaches 2, there are many such nodes. Therefore, each of these nodes is reachable from each other, and the walk down the degree sequence can continue down to nodes of degree of the order of $\sqrt{N}$.

The paper [ALPH01] later attempts to estimate the number of nodes needed to be traversed in order to search a finite fraction of the nodes in the network. Since the average number of neighbors of a node reached through a link is $\kappa - 1$ (see Chapter 10), the number of second neighbors of a node of degree k is on average $k(\kappa - 1)$. For $2 < \gamma < 3$, κ diverges with the system size as $\kappa \sim K^{3-\gamma} \sim N^{\frac{3-\gamma}{\gamma-1}}$, whereas $\langle k\rangle$ is a constant. This leads to the estimate that the number of nodes that need to be visited in order to search the entire network is of order $N/\kappa \sim N^{2\frac{\gamma-2}{\gamma-1}}$. When $\gamma \approx 2$, this number is very small.

However, the above calculation can be misleading, which can be viewed as a precursor of the caveats using the mean-field approach. $k(\kappa - 1)$ is the *average* number of second neighbors of a node. However, this number is not the typical number of next nearest neighbors of a node. Rather, it is influenced mainly by nodes with a very high degree. For example, consider a case in which $N^{0.9}$ nodes can all lie within a distance of 2 of each other, whereas all other nodes only have a finite number of second neighbors.[1] In this case, the average number of second neighbors per node will be $(N^{0.9} \cdot N^{0.9} + N \cdot \mathcal{O}(1))/N = \mathcal{O}(N^{0.8})$. This may lead us to expect that by traversing $\mathcal{O}(N/N^{0.8}) = \mathcal{O}(N^{0.2})$ of the nodes, one can search through the entire network. However, in this case almost all nodes have only a finite number of second neighbors. Therefore, a finite fraction of the network must be traversed to search through any finite number of the nodes. The generality of this can be understood from the results in Chapter 6. Therefore, every K_l that is not a constant belongs to the core of the network at a layer l, for some finite l. The $l + 2$ layer still does not consist of $\mathcal{O}(N)$ nodes in the $N \to \infty$ limit. Therefore, to search most (or any finite fraction) of the network, $\mathcal{O}(N)$ nodes must be traversed. Exercise 18.1 discusses the actual fraction of nodes covered by this strategy. See also [MST06] for more details.

Despite the above discussion, in practical cases, the search by degrees can be performed well. Numerical results [ALPH01] indicate that only a small fraction of the nodes actually need be traversed to search a large fraction of the network. This may imply that for reasonable network sizes, the method may be of practical value, even though the search is not performed well in the limit of very large systems.

A disadvantage of the method suggested is that the information stored in each node is quite large, since the number of second neighbors of a node may be very large. However, this method is easily understood and is very intuitive and simple. Note that methods for searching and routing in networks with limited information have been studied intensively and are known as "compact routing schemes." The definition and description of these techniques are beyond the scope of this book. However, the interested reader is referred to [Cow01, PU89, TZ01] for the definition and general results, and to [CCD06, KFY04] for more recent results pertaining specifically to scale-free networks. Another issue is the amount of information needed at every node in order to navigate in a network. In this respect, scale-free networks seem harder to navigate compared to ER networks. For definitions and results, see [RGMS05, RMS05, STR05, TRS05].

18.3 Searching in networks using shortcuts

An interesting result for searching in spatial networks using shortcuts was introduced by Kleinberg [Kle00a, Kle00b]. The model presented there is very similar to small-world networks (Section 4.2), the difference being that instead of adding completely random links to the underlying lattice, links are added with the probability depending on the lattice distance. In this model every node is connected to its lattice neighbors

[1] Note that this may not be a valid network configuration. It is presented only as a numerical example.

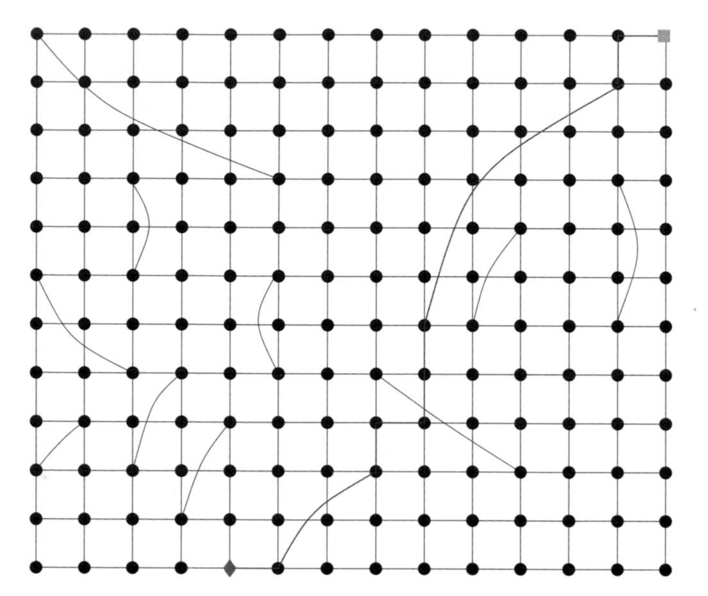

Figure 18.1 Illustration of the Kleinberg algorithm. A path goes from the source node (square) to the destination node (diamond). For simplicity, only a fraction of the shortcuts are shown.

and, in addition, one random link is added to each node,[2] i, with a probability $P(i, j) = c \cdot d(i, j)^{-\alpha}$ of connecting to a node, j, at a distance $d(i, j)$ from node i.

Kleinberg focuses on the two-dimensional case and on $\alpha = 2$, which is optimal in a sense that will be explained next. To find c, the normalization factor of the distribution, one should integrate over all possible destination nodes in the network,

$$1 = \sum_{j \neq i} c \cdot d(i, j)^{-\alpha} \approx \int_{r=1}^{\sqrt{N}} r^{-\alpha} \cdot 2\pi r \mathrm{d}r \approx c\pi \ln N \, . \tag{18.2}$$

Therefore, $c \approx (\pi \ln N)^{-1}$.

The search algorithm suggested by Kleinberg is very simple: from each node follow the link that brings you closest to the target node. See Figure 18.1 for an illustration. Since at every node encountered, there is at least one link bringing us closer to the destination, the distance to the destination always decreases, and therefore we arrive at a new node after every step we take. At every new node, we can treat the random link as if it is independent of the part of the network seen thus far.[3]

To analyze the behavior of the algorithm, one must partition the network into shells, where each shell, ℓ, contains all nodes at a distance at most 2^{ℓ} from the

[2] Adding random links with some constant probability at only a fraction of the nodes will give similar results.

[3] Since the path has gone through a very small portion of the network, its influence on the probability of a link leading to yet unvisited parts of the network is negligible.

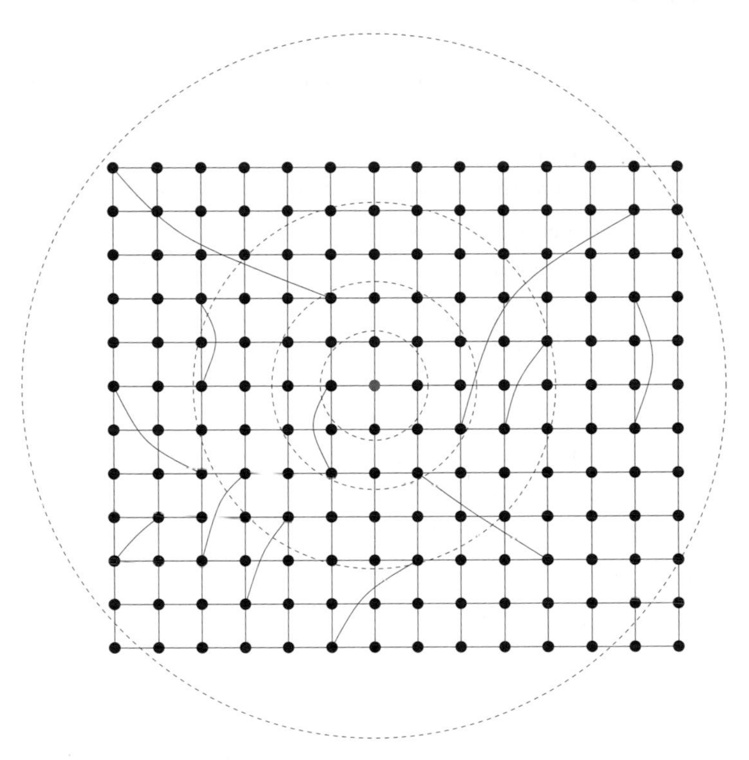

Figure 18.2 An illustration of Kleinberg's argument: partitioning the lattice into exponentially increasing regions. For simplicity the regions are shown as circles, although the distance on the lattice is the Manhattan distance (sum of distances in the x and y directions), which leads to diamond-shaped equidistance lines.

destination node (see Figure 18.2). The size of the entire lattice is $\sqrt{N} \times \sqrt{N}$, which indicates that the largest distance between nodes in the network is of the order of $\sqrt{N}$, meaning the maximum shell number is $\ell = \log_2 \sqrt{N} \sim \ln N$. The inner shells, for which $\ell \leq \ln \ln N$, have a diameter of the order of $\ln N$, and therefore any distance within them can always be traversed within $\ln N$ steps. If our current position is at some node in the ℓth shell, with $\ell > \ln \ln N$, all nodes within this shell are at a distance at most twice the radius of this shell, that is, at a distance of at most $2 \cdot 2^j$. This means that the probability of having a shortcut to a node within this shell is at least $c(2 \cdot 2^{\ell})^{-\alpha} \sim 2^{-2(\ell+1)}/\ln N$, for $\alpha = 2$. There are $\mathcal{O}\left(2^{\ell-1}\right)^2$ nodes within the $\ell - 1$ shell, so the probability of the next encountered shortcut to lead to *one* of the nodes in this regime is of order

$$(2^{\ell-1})^2 \frac{2^{-2(\ell+1)}}{\ln N} = \frac{1}{16 \ln N} . \tag{18.3}$$

Therefore, after approximately $m \equiv 16 \ln N$ steps, we encounter a shortcut leading us to an inner shell. If we do some multiple of m steps, the probability of not finding such a shortcut decays exponentially. Hence, after an order of $\ln N$ steps, we are bound to find a path to an inner shell. Since there are $\ln N$ shells, we will reach the destination in at most $\mathcal{O}(\ln^2 N)$ steps, much fewer than the $\sqrt{N}$ steps required in the original lattice.

Note that the ability to find such a short path results from the optimality of $\alpha = 2$. A larger α value would lead to a strong decay in the number of long-distance shortcuts, and therefore will require longer walks. A lower α will create many long-distance shortcuts that may lead us off the path. Although at the limit $\alpha \to 0$ a random graph is obtained, with distances of the order of $\ln N$, the network obtained is not easily searchable, since one may need to travel very far, geographically, to find a short path. Therefore, the underlying geography no longer provides information on the optimal path. A very similar result was obtained for modeling foraging (searching for food) in animals. For searching in a continuous two-dimensional case, it was determined that a Lévy flight with the probability of long jumps behaving as r^{-2}, with r being the distance, results in optimal searches [VBB+02, VBH+99].

For a general dimension d, Kleinberg obtains similar results, with $\alpha_{\text{opt}} = d$. The details are left as an exercise for the reader (Exercise 18.2). For a fractal, it was found that $\alpha_{\text{opt}} = d_{\text{f}}$ [Rb06]. The exact coefficients of the search time were obtained in [CCSb09].

18.4 Summary

Searching in complex networks of various classes can be made more efficient using the special properties of the networks. Furthermore, various network models may be better suited for different types of searches. As seen above, small-world networks with a certain dependence of the shortcut probability on the shortcut length lead to much faster searches than other networks. A recent, more elaborate study of this behavior can be found in [CCSb09]. This behavior is an example of the usefulness of the geographic information in searching for nodes. In different classes of networks, various types of hints, based on the label given to a node and using small routing tables, are helpful in quickly finding a short path between nodes. These methods are known as labeling schemes and compact routing, respectively. For more information see, for example, [Cow01, PU89, TZ01]. For information on fast navigating in complex networks see, for example, [BK09, CCD06, KFY04, RMS05, STR05].

An interesting question regarding the Milgram experiment is how did the participants manage to find a short path to the destinations. Even though a short path may exist it is not clear how it can be found given that people do not have the full map of the network, and are also not equipped with hints similar to those in compact routing schemes. In [WDN02] it is argued that such searches may be conducted using information on the nature of neighbor nodes. It is suggested that people possess a collection of properties, such as geographical residence area, occupation, areas of interest, etc. People may use such hints in finding short paths to other people. Each participant may send the letter to a contact having some properties that are closer to the destination than those of the sender. It is shown in [WDN02] that using such a mechanism one can find a short path to the destination using only local information, assuming some structure exists in the correlations between the linking probability and the properties of nodes.

Exercises

18.1 Using the generating function method (Chapter 9), calculate the probability of a node having a second neighbor (next nearest neighbor) of degree at least $k_{\max}$. Deduce what is the fraction of nodes that can be seen by traversing all nodes of degree at least $k_{\max}$ if every node has information on all its first and second neighbors.

18.2 Generalize the results in Section 18.3 to a lattice with a general dimension d.

18.3 What will be the result of the Kleinberg α_{opt} when only horizontal or vertical shortcuts are allowed?

19 Biological networks and network motifs

One of the most important classes of network studied in the literature is biological networks. This class contains a large variety of naturally occurring networks, formed by the long course of evolution. The human (and animal) body contains a large number of networks, some of which occur in real space, such as the networks of blood vessels, the bronchi, and the nerve system. These networks have been studied for a long time, and many of them are known to be fractal objects, see, for example [BH94, BH96, WBE97]. Another class of network, on which we will focus in this chapter, is logical. These are the networks of gene–gene, gene–protein, and protein–protein interactions. A survey of networks in biology can be found, for example, in [BO04].

In the cell, the genetic information is stored in DNA strands sitting in the cell's nucleus. To pronounce the genetic information DNA is transcribed (copied) to messenger RNA by a protein called RNA polymerase. The messenger RNA is then translated by ribosomes, which form amino acids according to the information in the RNA. Every three letter sequence represents a single amino acid from the twenty amino acids existing in humans and most other known organisms. A chain of amino acids is then created from a sequence of letters and forms a protein. The protein folds into a minimal energy configuration, whose shape determines most of its biological function. The created proteins then interact with the genetic information in several ways. A protein can connect to a location (called a promoter) in the DNA and suppress or promote the transcription of a sequence. This sequence may include the instruction for forming any other protein or itself. The entire transcription mechanism is controlled by proteins, including the ribosomes themselves which are large protein complexes.

Several possible network descriptions arise from the above scenario. A bipartite gene–protein network can be discussed, where proteins and genes are represented by two types of nodes, and are connected by a link whenever one affects the production or expression of the other. In a genetic network the nodes are genes and the links are between genes whose transcription affects the expression of each other. Metabolic (protein) networks are also studied, where proteins are linked if one affects the production of the other, or otherwise if they interact directly (chemically).

19.1 Structure of metabolic networks

Metabolic networks are networks of proteins, interacting with each other inside the cell. This is a directed network, since each protein may catalyze or repress the creation of every other protein, which does not necessarily imply the reverse process. In [JTA+00], the large-scale structure of metabolic networks was studied. It was concluded that for many species the metabolic network is scale free. Furthermore, it was observed that the diameter of the network is quite small (approximately 2–4 in most observed networks) and is practically independent of the network size. This independence may be explained by the results in Chapter 6 that indicate that the dependence of the diameter on N is very weak.

19.2 Structure of genetic networks

In genetic networks the nodes represent genes, and the directed links represent the influence of a gene on the pronunciation of another gene. Since each such phase requires a new generation of bacteria, which can take several minutes, these networks tend to be almost acyclic (i.e., there is a preferred direction towards which almost all links point) and shallow, with an average depth of about 5 for the well studied *E. coli* network. A survey of the structure of this network can be found in [MSOI+02].

19.3 Network motifs

Motifs are small building blocks of large networks. In this book we have mainly focused on the large-scale structure of complex networks. Small size features, such as triangles, have been treated as a product of geographical structure or as clustering phenomena, connected with some tunable parameter. However, it is widely believed that these features play a very important role in the function of the network. While this structure has many implications for functionality in terms of robustness, error tolerance etc., it does not supply information on the simple tasks that a network is supposed to perform.

In [IMK+03, MSOI+02] the structure of genetic networks was studied. The large-scale structure of these networks was found to be scale free. However, the small-scale structure seems non-random. The clustering in these networks is high. Looking more closely at the structure reveals that clustering is not enough to describe the

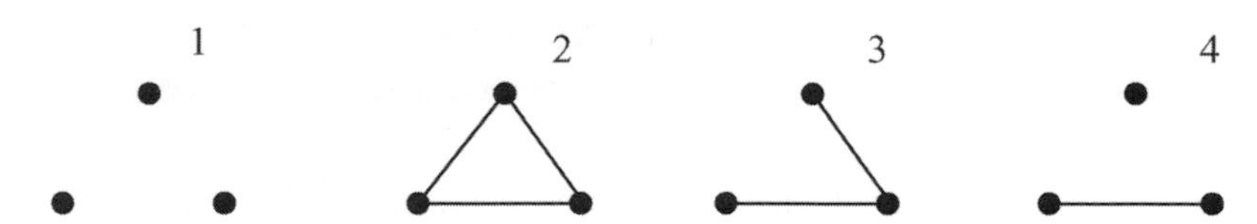

Figure 19.1 Four possible motifs of three nodes in an undirected network.

small-scale structure. When looking at small groups of nodes, one can distinguish between several network motifs. Motifs are the interconnection patterns of a group of nodes. When looking at three nodes in a graph they can be disconnected, form a path, a triangle, or they can consist of a connected pair and an unconnected node (see Figure 19.1). When the links are directed there are 13 possible motifs, while for four nodes there are already 230 possible motifs.

To check whether these motifs may have appeared by chance, the probability of such an appearance is estimated [IMK$^+$03]. In general, in a sparse directed ER network, $G_{N,p}$, the expected number of appearances of a subgraph, F, with n nodes and ℓ links is

$$N_F = \binom{N}{n} p^{\ell}(1-p)^{n(n-1)-\ell}\frac{n!}{a_0}\,, \tag{19.1}$$

where a_0 is the size of the automorphism group of F, i.e., the number of possible rearrangements of the nodes of F that leave its edges unchanged (see, for example, [AB02]). Notice that the probabilities for the appearance of the subgraph in non-disjoint sets of nodes is not independent. However, this is a good approximation when the graph is sparse. From Eq. (19.1) for sparse ER graphs, $N_F \sim \langle k\rangle^{\ell}\, N^{n-\ell}$. This result indicates that the probability of finding dense subgraphs (denser than loops, for instance, cliques) in sparse random graphs is vanishingly small. One should, however, take into account the effect of the degree distribution on the probability of finding a subgraph.

When approximating the expected number of subgraphs in scale-free networks, the effect of hubs should be taken into account. We will follow the example given in [IMK$^+$03]. The motif for the estimation appears in Figure 19.2. The probability of finding such a motif in a random directed network depends on the *in-*, *out-* and *mutual* (bidirectional) degrees of the nodes of the motif. The probability of having a directed link from node 1, with out-degree K_1, to node 2, with in-degree R_2, is $P(\text{edge1}) = K_1 R_2/N\langle k\rangle$, assuming that $K_1 R_2 \ll N\langle k\rangle$. The probability of having the second edge, given the existence of the first one, is $P(\text{edge2}|\text{edge1}) = (K_1-1)R_3/N\langle k\rangle$. The expected number of subgraphs with n nodes, g_{a} directed edges and g_{m} mutual edges is,

$$N_F = \frac{sN^{n-g_{\text{a}}-g_{\text{m}}}}{\langle K\rangle^{g_{\text{a}}}\langle M\rangle^{g_{\text{m}}}}\left\langle \prod_{j=1}^{n}\binom{K_{\sigma_j}}{k_j}\binom{R_{\sigma_j}}{r_j}\binom{M_{\sigma_j}}{m_j}\right\rangle_{\{\sigma\}}\,, \tag{19.2}$$

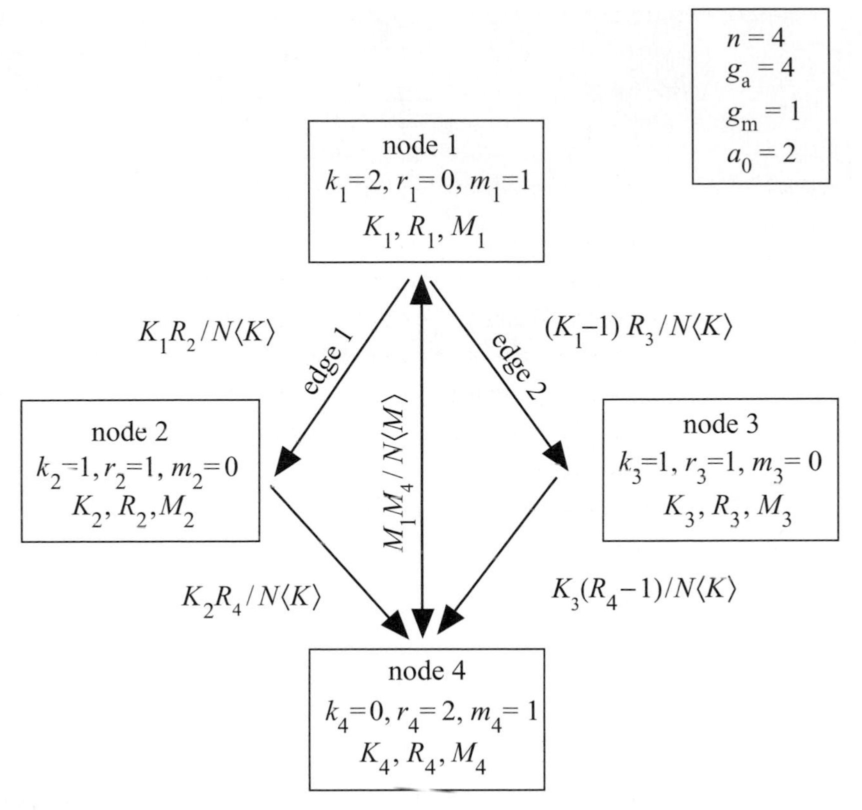

Figure 19.2 A subgraph with one mutual edge and four single edges. The subgraph degree sequences $\{k_i, r_i, m_i\}$ and node degrees $\{K_i, R_i, M_i\}$ are displayed in the boxes. Edge probabilities are displayed along the edges. Using Eq. (19.3), the mean subgraph number of appearances in an ensemble of random networks is $N_F = 2\langle K(K-1)M\rangle \langle R(R-1)M\rangle \langle RK\rangle^2 / N\langle K\rangle^4 \langle M\rangle$. After [IMK+03].

where $\{\sigma_1, \ldots, \sigma_n\}$ is a permutation of the nodes of the motif, and $s = a_0^{-1} \prod_{j=1}^{n} k_j! r_j! m_j!$ is the appropriate factor. The average in (19.2) reduces to a product of moments of different orders of the in-degree, out-degree and mutual degree distributions:

$$N_F = \frac{a N^{n-g_\mathrm{a}-g_\mathrm{m}}}{\langle K\rangle^{g_\mathrm{a}} \langle M\rangle^{g_\mathrm{m}}} \prod_{j=1}^{n} \left\langle \binom{K_i}{k_j}\binom{R_i}{r_j}\binom{M_i}{m_j}\right\rangle_i , \tag{19.3}$$

where the fact that each node should participate in the summation of only one term j introduces higher order corrections which we neglect. For example, subgraph id102 (Table 19.1), has $n = 3$ nodes, $g_\mathrm{a} = 2$ single edges and $g_\mathrm{m} = 1$ mutual edge. The subgraph degree sequences are $k_j = \{1, 1, 0\}, r_j = \{0, 1, 1\}$, and $m_j = \{1, 0, 1\}$.

Table 19.1 Mean numbers of the thirteen connected directed subgraphs in an ensemble of random networks with a given degree distribution

The degree distributions are those of transcription in the yeast *S. cerevisiae* ([MSOI+02]), synaptic connections between neurons in *C. elegans* ([WSTB86]), and World Wide Web hyperlinks between webpages in a single domain ([AJB99]). The theoretical values (Eq. (19.3)) are shown. Values below 0.5 were rounded to zero. In subgraphs marked with *, the theoretical values shown were obtained using the corrections for non-sparse graphs (see Appendix B of [IMK+03]). Subgraph id is determined by concatenating the rows of the subgraph adjacency matrix and representing the resulting vector as a binary number. The id is the minimal number obtained from all the isomorphic versions of the subgraph. After [IMK+03].

Subgraph id	Equation	Transcription	Neurons	WWW
6*	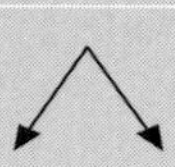 $N\langle K(K-1)\rangle/2$	1.2×10^4	4.3×10^2	4.7×10^7
12*	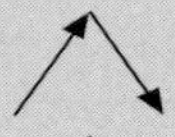 $N\langle KR\rangle$	3.6×10^2	8.7×10^2	2.5×10^6
14*	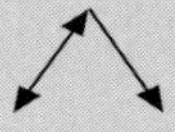 $N\langle KM\rangle$	1.9×10^1	8.7×10^1	3.8×10^6
36*	 $N\langle R(R-1)\rangle/2$	9.6×10^2	6.0×10^3	2.2×10^8
38	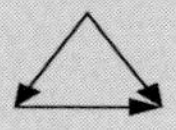 $\langle K(K-1)\rangle\langle RK\rangle\langle R(R-1)\rangle/\langle K\rangle^3$	1.3×10^1	1.2×10^2	3.4×10^5
46	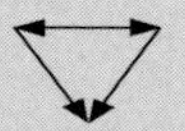 $\langle KM\rangle^2 \langle R(R-1)\rangle/2\langle K\rangle^2\langle M\rangle$	0	9.3	8.5×10^3
74*	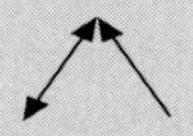 $N\langle RM\rangle$	2.9	1.3×10^2	4.8×10^6
78*	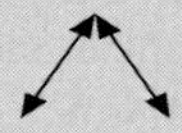 $N\langle M(M-1)\rangle/2$	0	6.6	2.5×10^7
98	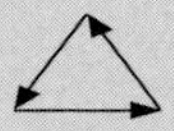 $\langle KR\rangle^3/3\langle K\rangle^3$	0	4.5	3.3×10^1
102	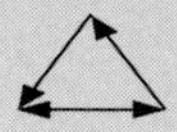 $\langle KM\rangle\langle RM\rangle\langle RK\rangle/\langle K\rangle^2\langle M\rangle$	0	2	1.4×10^2
108	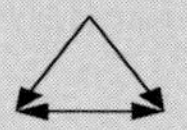 $\langle RM\rangle^2\langle K(K-1)\rangle/2\langle K\rangle^2\langle M\rangle$	0	1.4	2.9×10^3
110	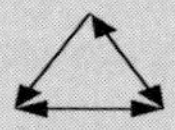 $\langle KM\rangle\langle RM\rangle\langle M(M-1)\rangle/\langle K\rangle\langle M\rangle^2$	0	0	2.3×10^3
238	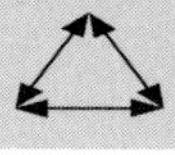 $\langle M(M-1)\rangle^3/6\langle M\rangle^3$	0	0	5×10^4

Using (19.3) we find:

$$N_F = \langle \text{id}102\rangle = \frac{\langle KM\rangle\langle RM\rangle\langle RK\rangle}{\langle K\rangle^2\langle M\rangle}\,. \tag{19.4}$$

When looking at real-world networks it can be observed that some motifs are much more pronounced than expected for random networks with the same characteristics.

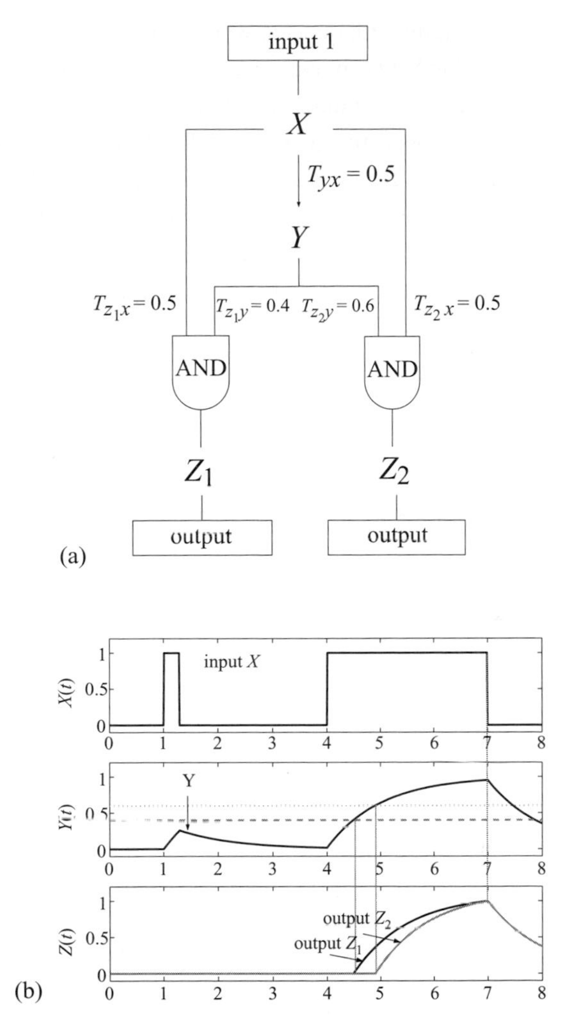

Figure 19.3 Kinetics of a double-output feed forward loop generalization following pulses of stimuli. (a) A double-output feed forward loop with positive regulation and AND-logic input function for Z_1 and Z_2. Numbers on the arrows are activation thresholds. (b) Simulated kinetics of the double-output feed forward loop in response to a short pulse and a long pulse of X activity. The dashed and dotted horizontal lines represent the activation thresholds T_{z_1y} and T_{z_2y}. After [KIMA04].

This has led to the understanding that these motifs play some important role in the functioning of the network. For instance, in genetic networks the feed-forward loop motif (number 38 in Table 19.1) is highly pronounced. This is attributed to the fact that this motif can be used as a filter for short pulses (see Figure 19.3). Such filters are important for filtering short disturbances in the outside environment, which should not be treated unless they prevail [SOMMA02]. In the WWW, motifs with mutual edges are highly pronounced, which is usually attributed to the sociological phenomena of linking between pages with related material or interest.

As a precaution, one should notice that when testing whether a motif is highly pronounced, the question of the initial model selection is highly important (for some debate over this issue see [ARFBTS04] and [MIK$^+$04]). For example, when a network is built from small communities [PN03] or has geographical underlying structure [IA05], many of the mentioned motifs may appear with high probability, when they are not expected to appear in randomly connected networks. Therefore, some care should be taken before it can be concluded that frequently occurring motifs have a functional role.

19.4 Summary

We have seen that in addition to the large-scale structure of networks, the small-scale building blocks may also be important in the functionality of networks. They may contain information on basic operations or the underlying structure of networks. In biology, evolution acts as a catalyst for the formation of simple units that can perform simple important functions in cells and organisms. It is not clear yet whether the large-scale scale-free topology of many biological networks is the result of selection processes or the result of the process creating these networks by mutations.

Some attempts to identify the process of network creation from the evolutionary data have also been conducted. Indications show that the evolutionary process may indeed be a Barabási–Albert process with linear preferential attachment. For results in this direction see [EL03].

A Appendix A Probability theoretical methods

In this appendix we will briefly describe probability theory and methods that will be used throughout the book. For a more complete treatment of probability theory, see any standard textbook on probability theory, such as [Fel68].

A.1 Probabilities and distributions

The main concept in probability theory is that of a probability space or an ensemble. The probability space consists of different objects or events, each of which has a measure – **probability** – assigned to it.

When each event is assigned a number, this number is termed a **random variable**. The random variable can be either discrete, in which case the measure is termed a **probability distribution**, or continuous, in which case the measure is termed "**probability density**." The single events then have an extremely small probability of occurring and only the probability of a range of values, obtained through integration assuming that the probability density is finite.

When a finite number of events exists, say N events, the most common case is to assume they are equiprobable. Therefore, the probability of finding some property is $P = M/N$, where M is the number of events having the property. If $P = 1 - f(N)$, where $\lim_{N\to\infty} f(N) = 0$, (i.e., the probability of finding the property approaches 1, when N becomes large) the property is said to hold **almost always** (or a.a.). In this book, when a property is said to hold, it actually means that it almost always holds.

When N independent trials are made, each of which has probability p of "success," the distribution of the number of successes, k, is

$$P(k) = \binom{N}{k} p^k (1-p)^{N-k} , \qquad \text{(A.1)}$$

where $\binom{k}{N} = N!/k!(N-k)!$. This is known as the **binomial distribution**.

If an event can occur at any time, with an average of λ events occurring, the distribution of the number of occurring events, k, is

$$P(k) = \frac{e^{-\lambda}\lambda^k}{k!} . \quad \text{(A.2)}$$

This is known as the **Poisson distribution**. The Poisson distribution also results from a binomial process where $N \to \infty$, and $Np \to \lambda$.

The **expectation** or **ensemble average** of a random variable x is the average of the random variables of all realizations. It will be denoted by $\langle x \rangle$, and is calculated using the equation

$$\langle x \rangle = \sum_x x P(x) . \quad \text{(A.3)}$$

Similarly, the **second moment** is defined by

$$\langle x^2 \rangle = \sum_x x^2 P(x) . \quad \text{(A.4)}$$

The **variance** is $V = \langle x^2 \rangle - \langle x \rangle^2$, and the **standard deviation** is $\sigma = \sqrt{V}$.

A.2 Continuous variables and probability densities

If a random variable, X, can obtain a continuous range of values, then, strictly speaking, the probability of obtaining any single value (except possibly a vanishingly small set) is extremely small. To handle such variables, one can use two possible probability measures: (1) the cumulative distribution function (CDF), $P(X > x_0)$, and (2) the probability density function (PDF), $p(X)$. The CDF gives the probability of having a result smaller than a given value. It should yield $P(X > x_0) = 0$ for any x_0 lower than the lowest possible value X can accept, $P(X > x_0) = 1$ for any x_0 higher than the highest possible value X can accept, and $0 \leq P(X > x_0) \leq 1$ for any x_0 in between. The PDF, $p(x_0)$, is actually not a probability, but rather a density function. It can obtain values higher than 1. It can be thought of as the density of the probability of obtaining a value in a small range, i.e., the probability of obtaining a value between x_0 and $x_0 + \Delta x$ is approximately $p(x_0)\Delta x$. The PDF must be non-negative, and obey the normalization condition:

$$\int_{X_{\min}}^{X_{\max}} p(X)\mathrm{d}X = 1 . \quad \text{(A.5)}$$

The PDF and CDF are related through the connection

$$P(X > x_0) = \int_{X_{\min}}^{x_0} p(X)\mathrm{d}X . \quad \text{(A.6)}$$

If one wishes to use a different random variable, Y, related to X by a one-to-one continuous correspondence, $Y(X)$, the PDFs are related by

$$|p(Y)\mathrm{d}Y| = |p(X)\mathrm{d}X| \,, \tag{A.7}$$

leading to

$$p(Y) = p(X)\left|\frac{\mathrm{d}X}{\mathrm{d}Y}\right| \,. \tag{A.8}$$

If the function is not one-to-one, then the contributions from all the preimages of Y should be summed.

The expectation and other moments of continuous distributions are defined similarly to discrete distributions according to

$$\langle X\rangle = \int Xp(X)\mathrm{d}X \,, \tag{A.9}$$

and, for every n,

$$\langle X^n\rangle = \int X^n p(X)\mathrm{d}X \,. \tag{A.10}$$

A.3 Conditional probabilities and Bayes' rule

If two (or more) events can occur randomly, one can define the conditional probability for event A given B,

$$P(A|B) = \frac{P(A, B)}{P(B)} \,, \tag{A.11}$$

where $P(A, B)$ is the probability of both A and B to occur together. The conditional probability is the probability of B to occur if it is known that A occurred.

From Eq. (A.11), it follows that $P(A, B) = P(B)P(A|B)$, but also $P(A, B) = P(A)P(B|A)$. From this, Bayes' rule follows

$$P(A|B) = \frac{P(B|A)P(A)}{P(B)} \,, \tag{A.12}$$

which allows us to swap the dependent and independent events, and is the basis for statistical inference.

A.4 Branching processes

Branching processes are processes where in each discrete generation every individual of the previous generation may give birth to a random number of individuals, independent of its siblings. The number of offspring of each individual, k, is distributed according to a distribution $P(k)$. The average number of offspring is $\langle k \rangle$, defined by $\langle k \rangle \equiv \sum kP(k)$. The average number of offspring after l generations is $\langle k \rangle^l$, which is easily seen to decay exponentially if $\langle k \rangle < 1$, and increase exponentially if $\langle k \rangle > 1$. If $P(0) \neq 0$, then the process has a finite probability of dying out after a finite number of generations. If $\langle k \rangle \leq 1$, the process will die out with probability 1 after a finite number of steps. If $\langle k \rangle > 1$, there is some probability $p > 0$ that the process will continue infinitely. A very efficient method for describing branching processes is the generating function method, described in the next section.

In some cases, more than one type of individual can exist in each generation. These cases are usually described by the more general framework of Markov processes (for continuous time), or Markov chains (for discrete time). If a finite number of individual types exists, the number of individuals can be described by a vector $\vec{N}_l$,

$$\vec{N}_l = \begin{pmatrix} n_1 \\ n_2 \\ n_3 \\ \vdots \end{pmatrix} . \quad \text{(A.13)}$$

The process can then be described by a transition matrix. If the probability of an individual of type a giving birth to k individuals of type b is $P_{ab}(k)$, then the transition matrix will be:

$$\mathbf{T} = \begin{pmatrix} \sum kP_{aa}(k) & \sum kP_{ba}(k) & \\ \sum kP_{ab}(k) & \sum kP_{bb}(k) & \\ & & \ddots \end{pmatrix} . \quad \text{(A.14)}$$

The survival probability is then determined by the largest eigenvalue of the matrix:

$$\lambda_{\max} \equiv \max_{\vec{v}} \frac{\mathbf{T}\vec{v}}{||\vec{v}||} . \quad \text{(A.15)}$$

If $\lambda_{\max} \leq 1$, the process will die out with probability 1, otherwise it has a finite survival probability.

A.5 Generating functions

The generating functions method (also known as the z-transform) can be used to solve exactly or approximately many problems in combinatorics and probability theory (for a comprehensive review, see [Wil94]). The main idea is, when given a sequence of values, $\{a_i\}$, which we wish to calculate, we define a power series:

$$A(x) = \sum_i a_i x^i \ . \tag{A.16}$$

At the moment, this power series is considered only as a formal one, without any consideration of convergence or any other analytic property.

A.5.1 Abelian and Tauberian theorems

One of the advantages of the generating function formalism is the possibility of obtaining asymptotic expansion of a formula in cases where an exact solution is impossible or unnecessary. To obtain this information, one should treat the power series as the expansion of a complex function. In this case, the Abelian and Tauberian theorems allow one to deduce the asymptotic behavior of the power series coefficients from the function's behavior near the critical point and vice versa. (For more detailed information see [Wei94, Wil94].)

The Abelian theorem asserts that if two power series have the same asymptotic behavior of the coefficients, then they will converge similarly at the critical point. For example, if the coefficients of the power series behave as $a_n \sim n^{-\gamma}$ (a common case in this book), with $0 < \gamma < 1$, then, regardless of the exact form of the coefficients, the series will diverge similarly. The behavior can be predicted by turning the sum:

$$G(z) = \sum_{n=0}^{\infty} a_n z^n \ , \tag{A.17}$$

into an integral, near the critical point, $z_\mathrm{c} = 1$. Taking $z = 1 - \epsilon \approx \mathrm{e}^{-\epsilon}$, the integral can evaluated as

$$G(1-\epsilon) \approx \int_0^\infty n^{-\gamma} \mathrm{e}^{-\epsilon n} \mathrm{d}n \sim \epsilon^{\gamma-1} \int_0^\infty x^{-\gamma} \mathrm{e}^{-x} \mathrm{d}x = \epsilon^{\gamma-1} \Gamma(1-\gamma) \ , \tag{A.18}$$

with higher order corrections. If $\gamma > 1$, the series converges for $z = 1$. However, repeated integration can be applied until the coefficients are in the range $0 < \gamma < 1$, and analytic terms will precede the term $\epsilon^{\gamma-1}$.

The Tauberian theorem allows the deduction in the opposite direction. If the behavior near the critical point is known, and some conditions are fulfilled (in particular, it should be known that the coefficients are monotonic, at least asymptotically), then it is possible to deduce the asymptotic behavior of the power series coefficients. The asymptotic expansion can be written as:

$$a_n \sim c\mathrm{e}^{-kn} n^{-\gamma} , \tag{A.19}$$

where k is determined by the radius of convergence of the function, R, according to the rule $k = \ln R$, and γ is determined by a process similar to Eq. (A.18).

Exercises

A.1 Find the expectation and variance of the Poisson and binomial distributions.

A.2 Find the z-transform of the Poisson and binomial distributions.

A.3 For what value does the Poisson distribution give a maximum?

A.4 If x is distributed uniformly between 1 and 2, what is the probability density of
(a) x^2,
(b) $\ln x$,
(c) e^x?

B Appendix B Asymptotics and orders of magnitude

Throughout the book we use the order of magnitude expression both in the discussion of algorithm complexity and in the discussion of the asymptotic behavior of expressions. Several symbols are used for order of magnitude notation. The most common one is known as **Big-O**, and the use of this expression is sometimes referred to as *Big-O notation*.

The most commonly used symbol is $\mathcal{O}(N)$, when an expression is said to behave as $\mathcal{O}(N)$. This indicates that the expression grows asymptotically linearly with N. Formally, this actually indicates that the expression grows *at most* linearly with N. Saying that $f(N) = \mathcal{O}(g(N))$, for some functions f and g formally means that

$$\lim_{N \to \infty} \frac{f(N)}{g(N)} < \infty \,. \tag{B.1}$$

Similarly, $f(N) = o(g(N))$ indicates that asymptotically $f(N)$ becomes infinitely smaller than $g(N)$ in the limit of large N values, i.e.,

$$\lim_{N \to \infty} \frac{f(N)}{g(N)} = 0 \,. \tag{B.2}$$

These symbols may also be used in other contexts, such as the limiting behavior of a function of x when $x \to 0$ and not $x \to \infty$, or at any other limit.

Less commonly used symbols are Ω, which is the converse of $\mathcal{O}$. That is $f(N) = \Omega(g(N))$ if $g(N) = \mathcal{O}(f(N))$, meaning that f grows at least as fast as g. Similarly, ω serves as the converse of o. Finally, Θ indicates that two functions behave similarly asymptotically, up to a constant factor, i.e., $f(N) = \Theta(g(N))$ if

$$|C'g(N)| < |f(N)| < |Cg(N)| \,, \tag{B.3}$$

for some positive $C, C' > 0$ and $N \to \infty$.

More information about Big-O notation in the context of computational complexity can be found, for example, in [CLR90].

Exercises

B.1 (a) Is $N = \mathcal{O}(N^2)$?
(b) Is $N^2 = \mathcal{O}(N)$?

(c) Is $N = \mathcal{O}(N)$?
(d) Is $N = o(N)$?

B.2 Order the following functions by order of magnitude at $N \to \infty$: $\ln N$, N, N^2, $N!$, e^N, N^{e}, $\ln N^{\ln N}$, $8N + 2\sqrt{N}$.

Appendix C Algorithms for network simulation and investigation

In this appendix we present efficient methods for simulating the networks discussed in the book and some efficient algorithms for analyzing their structure.

C.1 Simulation of generalized random graphs

To simulate general networks we have used the following algorithm.

(i) For each node choose a degree from the required distribution.
(ii) Create a list where each node is repeated as many times as the chosen degree.
(iii) Randomly choose pairs from the list and connect the chosen nodes (by storing for each node a list of neighbors, and adding each of the selected nodes node to the list of neighbors of the other node). Remove the chosen pair from the list (by replacing them with the last two entries in the list).

Double- and self-edges are ignored and if the number of entries in the list is odd (overall odd number of connections) one entry can also be discarded. This has a minor influence on the degree distribution if the network is large. Another possibility is to discard cases in which the sum of degrees is odd, and to rechoose the matching in case of a self loop or double-edge. Justification for the use of this algorithm for simulation and analysis as an algorithm generating the probability space of generalized random graphs can be found in [Bol80].

C.1.1 Generating scale-free networks

To generate degrees from a scale-free distribution we generate a random number, u, between 0 and 1 from a uniform distribution, and then generate a new number k using the formula:

$$k = \frac{m}{u^{1/(\gamma-1)}}. \tag{C.1}$$

This generates a random *real* number greater than m, with a distribution of $P(k) \propto k^{-\gamma}$. To prevent the appearance of overflows, u can be chosen such that it is large enough to yield numbers smaller than $N-1$ (since this is the maximum degree per node). If an upper cutoff is required a new random number is drawn whenever k is too large. After k is chosen, the closest integer is taken as the degree of the node. The resulting degree distribution is thus

$$P(k') = \int_{k'-1/2}^{k'+1/2} ck^{-\gamma} \mathrm{d}k, \tag{C.2}$$

where c is the normalization factor, except for the lower cutoff whose probability is given by:

$$P(m) = \int_{m}^{m+1/2} ck^{-\gamma} \mathrm{d}k. \tag{C.3}$$

This is in contrast to the probability chosen by [ACL00] and [CNSW00, NSW01], where a discrete scale-free distribution is used. The main difference is in the fraction of nodes of degree m (usually $m = 1$). Due to Eq. (C.3) the fraction of low-degree nodes in our derivation is lower and therefore the networks generated this way are more robust than those generated using the discrete distribution. However, the behavior of both distributions in the tail is approximately the same, and therefore the qualitative behavior of all phenomena influenced by the scale-free nature of the distribution should be the same.

The complexity of the algorithm depends upon the number of links in the network and therefore is of order $\mathcal{O}(N\langle k\rangle)$, which is fast enough to execute for large graphs. The space needed is again of order $\mathcal{O}(N\langle k\rangle)$. These limits allow the creation of networks of about 10^7 nodes.

C.1.2 Simulation of random breakdown

To simulate the process of random breakdown of a fraction p of the nodes, pN nodes are chosen at random. These nodes are removed and all their connections are also removed. The links are followed in order to remove all connections to the deleted nodes from the other end of the link.

After the removal of the nodes, the size of the spanning cluster (if it exists) and the finite clusters is measured. The measurement is performed using breadth first search from each node which has not been marked as probed. This method uncovers the graph cluster by cluster, and can also be used to find the distance (i.e., the shortest route) between a node on the cluster and all the other nodes on that cluster.

It should be noted that a very efficient, almost linear time algorithm for the investigation of the percolation threshold is presented in [NZ01]. This allows the study of percolation with link by link or node by node removal, where each removed link requires only $\mathcal{O}(1)$ operations (or sometimes slightly more).

C.1.3 Simulation of intentional attack

To simulate an intentional attack the same method as in random removal of nodes is applied. The only difference is that the removed nodes are not selected randomly, but are chosen as the highest degree nodes. Sorting the nodes would take $\mathcal{O}(N \ln N)$ operations, however, since no node has degree greater than N an array of lists of nodes for each degree can be produced, and the nodes can be removed starting from the highest degree downwards. This only requires $\mathcal{O}(N)$ operations, and therefore the entire execution of the program takes only $\mathcal{O}(N)$. A somewhat different result would be achieved if the list is updated every time a node is deleted, since this influences the degree of the other nodes. However, simulations show that this change has a small influence on the results, and the above analysis (in the previous section) is based on the static picture. Therefore, this is the method used to obtain the results given here.

C.2 Generating random directed networks

To generate a random directed network, an algorithm very similar to the standard algorithm for a generalized random network is used. For each node, the in- and out-degrees (j and k) are chosen from the distribution $P(j, k)$ and then two lists are created: one in which every node i is repeated j_i times, and one in which it is repeated k_i times. Then, a random member of each list is chosen, and a directed link is formed from the chosen member of the second list to the chosen member of the first. The process ends when the lists are empty. The condition $\langle j \rangle = \langle k \rangle$ ensures that the number of members of each list is approximately the same.

C.3 Networks with different properties: Metropolis style algorithms

In many classes of network, no simple method for the creation of random network of the class is known. In these cases a Monte Carlo method is usually invoked. These methods are based on selecting a random initial configuration, and then performing a series of probabilistic moves such that the distribution of networks obtained converges to the desired distribution. The method is based on the idea of a thermodynamic ensemble of network configurations.

The desired network class is determined by an energy function, which is defined on a network configuration. The function should be lower the more desirable its properties. For example, if one wishes to create networks with high clustering, the energy function may be the inverse of the clustering coefficient or the inverse of the number of triangles. A temperature is also chosen, where eventually, the probability of being at some configuration X approaches $P(X) = \exp(-E(X)/T)/z$, where $E(X)$ is the energy of the configuration, T is the temperature and z is an appropriate normalization factor. The temperature can be kept constant or can change during the execution of the algorithm.

In these algorithms one starts with a randomly chosen network, and then performs a (usually long) series of steps. In each such step the energy of the configuration is calculated, and then a manipulation is performed. The manipulation should be consistent with the requirements on the network class. For example, to move between all networks with the same number of links one can use deletion of a random link and the addition of another as a step. To conserve the degree sequence, one can choose two random links and switch between their destinations (i.e., remove the links (a, b) and (c, d) and add (a, d) and (c, b) instead). After the step is made, the energy function of the new configuration is calculated. Then the decision whether to stick with the new configuration or to return to the previous configuration is made with probability depending on the energy function. For good performance of the algorithm the energy function should be calculable locally. That is, the change in energy between configurations should depend only on the changed links, and possibly some near neighbors, so calculating the new energy should not require recalculating the energy for the entire network.

In order to work properly, the probabilities of switching between configurations, $P(X \to Y)$, should satisfy the detailed balance condition

$$P(X)P(X \to Y) = P(Y)P(Y \to X) . \tag{C.4}$$

Usually, for fast execution, one chooses $P(X \to Y) = 1$ if $E(X) > E(Y)$ and otherwise $P(X \to Y) = \mathrm{e}^{(E(Y)-E(X))/T}$. This condition still does not guarantee the convergence to the desired distribution. This also depends on the moves used. A path between any two configurations consisting only of the used manipulations should exist, and should preferably be short for fast convergence. Also, the same number of manipulations should be available from any configuration.

The Metropolis algorithm may take very long to converge. It may fall into a local minimum of the energy landscape and take an unreasonably long time to leave it. In order to increase the convergence probability it is sometimes attempted to start with a high temperature, making all moves probable and covering the entire configuration space, and gradually decrease the temperature in order to approach the largest minimum. This method is used commonly in searching algorithms in complicated optimization problems and in artificial intelligence, and is known as "simulated annealing." However, Metropolis type algorithms do not always converge to the desired distribution in reasonable time.

C.4 Finding shortest and optimal paths

Finding the shortest path between two nodes in a network is one of the most common algorithmic tasks, and many algorithms exist for this problem. For a survey of such algorithms, as well as many other useful algorithms see, for example, [CLR90].

The simplest method of finding the shortest path in terms of the hop count (i.e., when all links have the same weight) is using breadth first search (BFS). Breadth first search finds the shortest path between a single source node and every other node in the network. Breadth first search is based on starting at a node, whose distance is set to zero, and adding all its neighbors to a queue. The nodes in the queue are explored on a first in first out order, where each node that is reached is marked as explored. For each node explored, i, all of i's unexplored neighbors that are not yet in the queue are added to the queue, and their distance is set as $d_i + 1$ where d_i is the distance of node i from the initial node.

A more advanced algorithm, capable of handling weighted networks, is the Dijkstra algorithm. This algorithm works by dividing the nodes in the network into three different groups, say the white, gray and black groups. In the beginning all nodes are colored white, except for the origin, i, which is colored black and assigned a distance of zero. Then all of the origin's neighbors are colored gray and each of them, j, is assigned a distance of $d_{ij} = w_{ij}$, where w_{ij} is the weight of the link between i and j. Then, and at every next stage, the lowest distance gray node, j, is colored black and each of its neighbors, l, is colored gray and assigned a distance $d_{il} = d_{ij} + w_{jl}$. If node l was already colored gray, because it was reached before from another node, it is assigned the minimum of its current d_{il} and $d_{ij} + w_{jl}$. The algorithm ends when all nodes are colored black. This algorithm guarantees that no node is colored black until no possible shorter path exists for this node.

If the shortest paths between all pairs of nodes, rather than from one source to all nodes, is desired, the Floyd–Warshall algorithm can be used, and gives better performance than the application of Dijkstra's algorithm from all possible sources.

The optimal paths can be found in the same manner as the shortest path is found. If strong disorder is desired, the weights can be chosen from a very wide distribution (which may lead to numerical problems), or otherwise, an algorithm such as Dijkstra's can be used with a lexicographically ordered vector of weights serving as the distance function, and the lexicographic ordering serving as a distance comparison method instead of regular scalar comparison.

Another alternative in the case of strong disorder is to use the bombarding algorithm, where the links are removed in decreasing weight order, but a link is not removed if it disconnects a component in the network into two components. Similarly, one can start from a network with no links and add links in increasing weight order, but skipping links that create loops. In both cases the network of shortest paths is eventually a tree, which is also the minimum spanning tree of the original network.

C.5 The minimum spanning tree

A spanning tree is a subset of the links in the network which is a tree (graph with no loops) that spans all the nodes in the network. The minimum spanning tree is the spanning tree that has the minimal total weight among all spanning trees.

Several algorithms exist for finding the minimum spanning tree. The most well known are the Prim algorithm and the Kruskal algorithm (see, for example, [CLR90]). In the Prim algorithm one starts with a single node (the root) and adds to it the minimum weight link of this node and the node to which it leads. The two nodes are now the tree, and the algorithm continues by adding to the tree, at each step, the smallest weight link adjacent to it that does not form a loop. The process continues until all nodes are connected. (This is achieved after $N - 1$ links have been added.) Notice that this process is similar to the bombarding algorithm presented in Section C.4.

In the Kruskal algorithm one starts with the network being a collection of trees (a **forest**), where each node is a single tree with no links. At every step the link of minimal weight among all links connecting two distinct trees is added. The algorithm ends when only one tree remains.

C.6 Finding communities in networks

An active field in complex network research is the study of methods for finding communities in networks. Given a network, the goal is to divide the nodes of the network into different communities. Each community consists of nodes that have some common function of a strong relation between them. When no knowledge of the properties of the nodes is given, except for the network structure, it is desirable to deduce something about the different communities using just information on the network topology. The aim of community detection algorithms is to divide the nodes of the network such that nodes inside each community are strongly connected, while the connections between different communities are weak. These algorithms are based on dividing the network into different groups of nodes, where the number (or total weight) of links inside a group is relatively large, while the connections between different groups are relatively sparse.

One algorithm for finding communities in networks is superparamagnetic clustering [BWD96]. In this method each node, i, is augmented with a Potts spin, s_i, which is a spin which takes one of the q states $1, 2, \ldots, q$. The energy of the system is $E = -J_{ij} \sum_{i,j} \delta_{s_i,s_j}$. The δ_{s_i,s_j} is the Kronecker delta, giving 1 if $s_i = s_j$ and 0 otherwise, and J_{ij} is a coupling constant, representing the strength of interaction between nodes i and j. This can equal 1 if there is a link between i and j and 0 otherwise, or it can equal the weight of the link between i and j.

In the execution of the algorithm, the system starts at temperature 0 with all spins having the same value. Then, the temperature is gradually raised. At each temperature, a series of Monte Carlo steps is executed (see Section C.3). When the temperature is raised, clusters start forming where the nodes in each cluster have the same spin value, but different clusters have different spins. Raising the temperature higher results in the breaking of clusters into smaller clusters until, eventually, the whole network breaks into single node clusters. This algorithm defines a tree of clusters, where, at the root the whole network is a single cluster, and at each level each node belongs to some unique cluster. The lower the level in the tree, the more fine grained is the division into communities.

The main idea of this algorithm is that communities are strongly connected internally, while being relatively sparsely connected to other communities. Therefore, the interactions inside communities are strong, and spins inside communities tend to align in the same direction, whereas between communities interactions are weak. Thus, when raising the temperature, correlations between communities break before correlations inside communities.

Another approach, presented in [GN02], is based on consecutively removing links of high betweenness (see Section 3.3). Since links between different communities are sparse, all communication between communities is conducted through a small number of links, and therefore each of these links is expected to have high betweenness centrality. Inside communities, links are dense, and many paths exist between most nose pairs in a community and therefore these links have low betweenness and will be removed only at late stages. This algorithm also leads to a tree of clusters, similar to the superparamagnetic algorithm.

Another class of algorithms is based on eigenvectors of the Laplacian matrix [New06] (see Chapter 17).

References

[AB02] R. Albert and A.-L. Barabási, Statistical mechanics of complex networks, *Reviews of Modern Physics* **74** (2002), 47.

[ABR06] L. Addario-Berry, N. Broutin, and B. Reed, The diameter of the minimum spanning tree of a complete graph, *Proceedings of the Fourth Colloquium on Mathematics and Computer Science, DMTCS, Nancy, France*, 2006, pp. 237–248.

[ACL00] W. Aiello, F. Chung, and L. Lu, A random graph model for massive graphs, *Proceedings of the 32nd ACM Symposium on Theory of Computing*, 2000, pp. 171–180. New York: ACM.

[ACL01] W. Aiello, F. R. K. Chung, and L. Lu, Random evolution in massive graphs, *IEEE Symposium on Foundations of Computer Science*, 2001, pp. 510–519. New York: IEEE.

[AJB99] R. Albert, H. Jeong, and A.-L. Barabási, Diameter of the World Wide Web, *Nature* **401** (1999), 130.

[AJB00] R. Albert, H. Jeong, and A.-L. Barabási, Error and attack tolerance of complex networks, *Nature* **406** (2000), 378–382.

[AJB01] R. Albert, H. Jeong, and A.-L. Barabási, correction: error and attack tolerance of complex networks, *Nature* **409** (2001), 542.

[ALPH01] L. A. Adamic, R. M. Lukose, A. R. Puniyani, and B. A. Huberman, Search in power law networks, *Physical Review E* **64** (2001), 046135.

[AM92] R. M. Anderson and R. M. May, *Infectious Diseases in Humans: Dynamics and Control*, 1992. Oxford: Oxford University Press.

[Apo67] T. M. Apostol, *Calculus*, volume 1: *One-Variable Calculus with an Introduction to Linear Algebra*, second edition, 1967. Wiley.

[ARFBTS04] Y. Artzy-Randrup, S. J. Fleishman, N. Ben-Tal, and L. Stone, Comment on "Network motifs: simple building blocks of complex networks" and "Superfamilies of evolved and designed networks," *Science* **305** (2004), 1107.

[AS00] N. Alon and J. H. Spencer, *The Probabilistic Method*, Wiley-Interscience Series in Discrete Mathematics and Optimization, 2000. New York: Wiley-Interscience.

[ASBS00] L. A. N. Amaral, A. Scala, M. Barthelemy, and H. E. Stanley, Classes of small-world networks, *Proceedings of the National Academy of Sciences of the USA* **97** (2000), 11149–11152.

[BA99] A.-L. Barabási and R. Albert, Emergence of scaling in random networks, *Science* **286** (1999), 509–512.

[BAJ00] A.-L. Barabási, R. Albert, and H. Jeong, Scale-free characteristics of random networks: the topology of the World Wide Web, *Physica A* **281** (2000), 69–77.

[Bar03] A.-L. Barabási, *Linked: How Everything is Connected to Everything Else and What it Means for Business, Science, and Everyday Life*, 2003. New York: Plume.

[Bar04] M. Barthélemy, Betweenness centrality in large complex networks, *European Physical Journal B* **38** (2004), 163.

[Bax82] R. J. Baxter, *Exactly Solved Models in Statistical Mechanics*, 1982. London: Academic Press.

[BB01] G. Bianconi and A.-L. Barábasi, Bose–Einstein condensation in complex networks, *Physical Review Letters* **86** (2001), 5632–5635.

[BBC+03] L. A. Braunstein, S. V. Buldyrev, R. Cohen, S. Havlin, and H. E. Stanley, Optimal paths in disordered complex networks, *Physical Review Letters* **91** (2003), 168701.

[BBPV04] A. Barrat, M. Barthélemy, R. Pastor-Satorras, and A. Vespignani, The architecture of complex weighted networks, *Proceedings of the National Academy of Sciences of the USA* **101** (2004), 3747–3752.

[BBV08] A. Barrat, M. Barthélemy, and A. Vespignani, *Dynamical Processes on Complex Networks*, 2008. New York: Cambridge University Press.

[BCK01] Z. Burda, J. D. Curreira, and A. Krzywicki, Statistical ensemble of scale-free random graphs, *Physical Review E* **64** (2001), 046118.

[BCL+04] G. Bonanno, G. Caldarelli, F. Lillo, S. Miccichè, N. Vandewalle, and R. N. Mantegna, Networks of equities in financial markets, *European Physical Journal B* **38** (2004), 363–371.

[BF06] M. Barthelemy and A. Flammini, Optimal traffic networks, *Journal of Statistical Mechanics* (2006), L07002.

[BH94] A. Bunde and S. Havlin (eds.), *Fractals in Science*, 1994. Berlin: Springer.

[BH96] A. Bunde and S. Havlin (eds.), *Fractals and Disordered Systems*, 1996. Berlin: Springer.

[bH00] D. ben-Avraham and S. Havlin, *Diffusion and Reactions in Fractals and Disordered Systems*, 2000. Cambridge: Cambridge University Press.

[BJN+02] A.-L. Barabási, H. Jeong, Z. Neda, E. Ravasz, A. Schubert, and T. Vicsek, On the topology of the scientific collaboration networks, *Physica A* **311** (2002), 590–614.

[BK03] Z. Burda and A. Krzywicki, Uncorrelated random networks, *Physical Review E* **67** (2003), 046118.

[BK09] M. Boguñá and D. Krioukov, Navigating ultrasmall worlds in ultrashort time, *Physical Review Letters* **102** (2009), 058701.

[BKM+00] A. Broder, R. Kumar, F. Maghoul, P. Raghavan, S. Rajagopalan, R. Stata, A. Tomkins, and J. Wiener, Graph structure in the web, *Computer Networks* **33** (2000), 309.

[BLM+06] S. Boccaletti, V. Latora, Y. Moreno, M. Chavez, and D.-U. Hwang, Complex networks: structure and dynamics, *Physics Reports* **424** (2006), 175–308.

[BLW94] J. B. Bassingthwaighte, L. S. Liebovitch, and B. J. West, *Fractal Physiology*, 1994. New York: Oxford University. Press.

[BMS04] P. S. Bearman, J. Moody, and K. Stovel, Chains of affection: the structure of adolescent romantic and sexual networks, *American Journal of Sociology* **110** (2004), 44–91.

[BNKM01] E. Ben-Naim, P. L. Krapivsky, and S. N. Majumdar, Extremal properties of random trees, *Physical Review E* **64** (2001), 035101(R).

[BO04] A.-L. Barabási and Z. N. Oltvai, Network biology: understanding the cell's functional organization, *Nature Reviews in Genetics* **5** (2004), 101–113.

[Boc08] S. Boccaletti, *The Synchronized Dynamics of Complex Systems*, Monograph Series on Nonlinear Science and Complexity, volume 6, 2008. Amsterdam: Elsevier.

[Bol80] B. Bollobás, A probabilistic proof of an asymptotic formula for the number of labelled regular graphs, *European Journal of Combinatorics* **1** (1980), 311–316.

[Bol85] B. Bollobás, *Random Graphs*, 1985. London: Academic Press.

[BP94] A. Berman and R. J. Plemmons, *Nonnegative Matrices in the Mathematical Sciences*, 1994. Philadelphia, PA; SIAM.

[BP02] M. Boguñá and R. Pastor-Satorras, Epidemic spreading in correlated complex networks, *Physical Review E* **66** (2002), 047104.

[BPV03] M. Boguñá, R. Pastor-Satorras, and A. Vespignani, Absence of epidemic threshold in scale-free networks with degree correlations, *Physical Review Letters* **90** (2003), 028701.

[BR02] B. Bollobás and O. M. Riordan, Mathematical results on scale free random graphs. In *Handbook of Graphs and Networks*, S. Bornholdt and H. G. Schuster, eds., 2002. Berlin: Wiley-VCH.

[BR04] B. Bollobás and O. Riordan, Coupling scale-free and classical random graphs, *Internet Mathematics* **1** (2004), 215–225.

[bRCH03] D. ben-Avraham, A. F. Rozenfeld, R. Cohen, and S. Havlin, Geographical embedding of scale free networks, *Physica A* **330** (2003), 107–116.

[BS02] S. Bornholdt and H. G. Schuster (eds.), *Handbook of Graphs and Networks*, 2002. Berlin: Wiley-VCH.

[BTW87] P. Bak, C. Tang, and K. Wiesenfeld, Self-organized criticality – an explanation of 1/f noise, *Physical Review Letters* **59** (1987), 381–384.

[BWC$^+$07] L. A. Braunstein, Z. Wu, Y. Chen, S. V. Buldyrev, S. Sreenivasan, T. Kalisky, R. Cohen, E. Lopez, S. Havlin, and H. E. Stanley, Optimal path and minimal spanning trees in random weighted networks, *International Journal of Bifurcation and Chaos* **17** (2007), 2215–2255.

[BWD96] M. Blatt, S. Wiseman, and E. Domany, Superparamagnetic clustering of data, *Physical Review Letters* **76** (1996), 3251–3254.

[CbH02] R. Cohen, D. ben-Avraham, and S. Havlin, Percolation critical exponents in scale free networks, *Physical Review E* **66** (2002), 036113.

[CCD06] S. Carmi, R. Cohen, and D. Dolev, Searching complex networks efficiently with minimal information, *Europhysics Letters* **74** (2006), 1102–1108.

[CCGJ02] Q. Chen, H. Chang, R. Govindan, and S. Jamin, The origin of power laws in internet topologies revisited, *INFOCOM 2002. Proceedings of the Twenty-First Annual Joint Conference of the IEEE Computer and Communications Societies* **2** (2002), 608–617.

[CCSb09] S. Carmi, S. Carter, J. Sun, and D. ben-Avraham, Asymptotic behavior of the kleinberg model, *Physical Review Letters* **102** (2009), 238702.

[CEbH00] R. Cohen, K. Erez, D. ben-Avraham, and S. Havlin, Resilience of the Internet to random breakdown, *Physical Review Letters* **85** (2000), 4626–4628.

[CEbH01] R. Cohen, K. Erez, D. ben-Avraham, and S. Havlin, Breakdown of the Internet under intentional attack, *Physical Review Letters* **86** (2001), 3682–3685.

[CGA02] J. Camacho, R. Guimerà, and L. A. N. Amaral, Robust patterns in food web structure, *Physical Review Letters* **88** (2002), 228102.

[CH03] R. Cohen and S. Havlin, Scale free networks are ultrasmall, *Physical Review Letters* **90** (2003), 058701.

[CHb02] R. Cohen, S. Havlin, and D. ben-Avraham, Structural properties of scale free networks, In *Handbook of Graphs and Networks*, S. Bornholdt and H. G. Schuster, eds., 2002. Berlin: Wiley-VCH.

[CHb03] R. Cohen, S. Havlin, and D. ben-Avraham, Efficient immunization strategies of computer networks and populations, *Physical Review Letters* **91** (2003), 247901.

[CHK+01] D. S. Callaway, J. E. Hopcroft, J. M. Kleinberg, M. E. J. Newman, and S. H. Strogatz, Are randomly grown graphs really random?, *Physical Review E* **64** (2001), 041902.

[CHK+07] S. Carmi, S. Havlin, S. Kirkpatrick, Y. Shavitt, and E. Shir, From the cover: a model of Internet topology using k-shell decomposition, *Proceedings of the National Academy of Sciences of the USA* **104** (2007), 11150–11154.

[Chu97] F. R. K. Chung, *Spectral Graph Theory*, CBMS Regional Conference Series in Mathematics, no. 92, 1997. Providence, RI: American Mathematical Society.

[CL01] F. Chung and L. Lu, The diameter of sparse random graphs, *Advances in Applied Mathematics* **26** (2001), 257–279.

[CL02] F. Chung and L. Lu, The average distance in random graphs with given expected degrees, *Proceedings of the National Academy of Sciences of the USA* **99** (2002), 15879–15882.

[CLHS06] Y. Chen, E. Lopez, S. Havlin, and H. E. Stanley, Universal behavior of optimal paths in weighted networks with general disorder, *Physical Review Letters* **96** (2006), 068702.

[CLR90] T. H. Cormen, C. Leiserson, and R. Rivest, *Introduction to Algorithms*, 1990. Cambridge, MA: MIT Press.

[CM05] A. Clauset and C. Moore, Accuracy and scaling phenomena in Internet mapping, *Physical Review Letters* **94** (2005), 018701.

[CMB99] M. Cieplak, A. Maritan, and J. R. Banavar, Optimal paths and growth processes, *Physica A* **266** (1999), 291–298.

[CNSW00] D. S. Callaway, M. E. J. Newman, S. H. Strogatz, and D. J. Watts, Network robustness and fragility: percolation on random graphs, *Physical Review Letters* **85** (2000), 5468–5471.

[Con89] A. Coniglio, Fractal structure of ising and potts clusters: exact results, *Physical Review Letters* **62** (1989), 3054–3057.

[Cow01] L. Cowen, Compact routing with minimum stretch, *Journal of Algorithms* **38** (2001), 170–183.

[CPH+08] Y. Chen, G. Paul, S. Havlin, F. Liljeros, and H. E. Stanley, Finding a better immunization strategy, *Physical Review Letters* **101** (2008), 058701.

[CRS+03] R. Cohen, A. F. Rozenfeld, N. Schwartz, D. ben-Avraham, and S. Havlin, Directed and non-directed scale free networks, *Proceedings of the XVIII Sitges Conference*, R. Pastor-Satorras and J. M. Rubí, eds., Lecture Notes in Physics, volume 625, 2003, p. 23. Berlin: Springer.

[CS01] R. F. I. Cancho and R. V. Sole, The small world of human language, *Proceedings of the Royal Society of London, Series B, Biological Sciences* **268** (2001), 2261–2266.

[DB02] Z. Dezső and A.-L. Barabási, Halting viruses in scale free networks, *Physical Review E* **65** (2002), 055103(R).

[DC02] J. Dall and M. Christensen, Random geometric graphs, *Physical Review E* **66** (2002), 016121.

[DCG99] R. De Castro and J. W. Grossman, Famous trails to Paul Erdős, *Mathematical Intelligencer* **21** (1999), 51–63.

[DD01] R. Dobrin and P. M. Duxbury, Minimum spanning trees on random networks, *Physical Review Letters* **86** (2001), 5076–5079.

[DG08] S. N. Dorogovtsev and A. V. Goltsev, Critical phenomena in complex networks, *Reviews of Modern Physics* **80** (2008), 1275–1335.

[DGM02a] S. N. Dorogovtsev, A. V. Goltsev, and J. F. Mendes, Pseudofractal scale-free web, *Physical Review E* **65** (2002), 066122.

[DGM02b] S. N. Dorogovtsev, A. V. Goltsev, and J. F. F. Mendes, Ising model on networks with an arbitrary distribution of connections, *Physical Review E* **66** (2002), 016104.

[DGM06] S. N. Dorogovtsev, A. V. Goltsev, and J. F. F. Mendes, k-Core organization of complex networks, *Physical Review Letters* **96** (2006), 040601.

[DM01a] S. N. Dorogovtsev and J. F. F. Mendes, Comment on "breakdown of the Internet under intentional attack," *Physical Review Letters* **87** (2001), 219801.

[DM01b] S. N. Dorogovtsev and J. F. F. Mendes, Language as an evolving word web, *Proceedings of the Royal Society of London, Series B* **268** (2001), 2603.

[DM02] S. N. Dorogovtsev and J. F. F. Mendes, Evolution of networks, *Advances in Physics* **51** (2002), 1079.

[DM03] S. N. Dorogovtsev and J. F. F. Mendes, *Evolution of Networks: From Biological Nets to the Internet and WWW*, 2003. Oxford: Oxford University Press.

[DMO06] S. N. Dorogovtsev, J. F. F. Mendes, and J. G. Oliveira, Degree-dependent intervertex separation in complex networks, *Physical Review E* **73** (2006), 056122.

[DMS00] S. N. Dorogovtsev, J. F. F. Mendes, and A. N. Samukhin, Growing network with heritable connectivity of nodes, ArXiv Condensed Matter e-prints (2000).

[DMS01a] S. N. Dorogovtsev, J. F. F. Mendes, and A. N. Samukhin, Anomalous percolating properties of growing networks, *Physical Review E* **64** (2001), 066110.

[DMS01b] S. N. Dorogovtsev, J. F. F. Mendes, and A. N. Samukhin, Giant strongly connected component of directed networks, *Physical Review E* **64** (2001), 025101(R).

[DMS01c] S. N. Dorogovtsev, J. F. F. Mendes, and A. N. Samukhin, Size-dependent degree distribution of a scale-free growing network, *Physical Review E* **63** (2001), 062101.

[DMS03] S. N. Dorogovtsev, J. F. F. Mendes, and A. N. Samukhin, Metric structure of random networks, *Nuclear Physics B* **653** (2003), 307.

[Doy02] J. P. Doye, Network topology of a potential energy landscape: a static scale-free network, *Physical Review Letters* **88** (2002), 238701.

[dSP65] D. J. de Solla Price, Networks of scientific papers, *Science* **149** (1965), 510–515.

[DYB03] G. F. Davis, M. Yoo, and W. E. Baker, The small world of the American corporate elite, 1982–2001, *Strategic Organization* **1** (2003), 301–326.

[ECC$^+$05] V. M. Eguiluz, D. R. Chialvo, G. A. Cecchi, M. Baliki, and A. V. Apkarian, Scale-free brain functional networks, *Physical Review Letters* **94** (2005), 018102.

[EL03] E. Eisenberg and E. Y. Levanon, Preferential attachment in the protein network evolution, *Physical Review Letters* **91** (2003), 138701.

[EMB02] H. Ebel, L.-I. Mielsch, and S. Bornholdt, Scale-free topology of e-mail networks, *Physical Review E* **66** (2002), 035103(R).

[ER59] P. Erdős and A. Rényi, On random graphs, *Publicationes Mathematicae* **6** (1959), 290–297.

[ER60] P. Erdős and A. Rényi, On the evolution of random graphs, *Publications of the Mathematical Institute of the Hungarian Academy of Sciences* **5** (1960), 17–61.

[ER61] P. Erdős and A. Rényi, On the strength of connectedness of a random graph, *Acta Mathematica Scientia Hungary* **12** (1961), 261–267.

[FDBV01] I. J. Farkas, I. Derényi, A.-L. Barabási, and T. Vicsek, Spectra of "real-world" graphs: beyond the semicircle law, *Physical Review E* **64** (2001), 026704.

[Fed88] J. Feder, *Fractals*, 1988. New York: Plenum Press.

[Fei97] U. Feige, Collecting coupons on trees, and the cover time of random walks, *Computational Complexity* **6** (1997), 341–356.

[Fel68] W. Feller, *An Introduction to Probability Theory and its Applications*, volume 1, 1968. New York: Wiley.

[Fel81] S. L. Feld, The focused organization of social ties, *American Journal of Sociology* **86** (1981), 1015.

[FFF99] M. Faloutsos, P. Faloutsos, and C. Faloutsos, On power-law relationships of the Internet topology, *Computer Communication Review* **29** (1999), 251.

[FHL01] P. Frojdh, M. Howard, and K. B. Lauritsen, Directed percolation and other systems with absorbing states: Impact of boundaries, *International Journal of Modern Physics B* **15** (2001), 1761.

[FiCJS01] R. Ferrer i Cancho, C. Janssen, and R. V. Solé, Topology of technology graphs: small world patterns in electronic circuits, *Physical Review E* **64** (2001), no. 4, 046119.

[Flo41] P. J. Flory, Molecular size distribution in three dimensional polymers. iii. tetrafunctional branching units, *Journal of the American Chemical Society* **63** (1941), 3096–3100.

[GA04] L. K. Gallos and P. Argyrakis, Absence of kinetic effects in reaction-diffusion processes in scale-free networks, *Physical Review Letters* **92** (2004), 138301.

[GAH08] A. Garas, P. Argyrakis, and S. Havlin, The structural role of weak and strong links in a financial market network, *European Physical Journal B* **63** (2008), 265–271.

[GCA+05] L. K. Gallos, R. Cohen, P. Argyrakis, A. Bunde, and S. Havlin, Stability and topology of scale-free networks under attack and defense strategies, *Physical Review Letters* **94** (2005), 188701.

[GCV+07] K.-I. Goh, M. E. Cusick, D. Valle, B. Childs, M. Vidal, and A.-L. Barabási, The human disease network, *Proceedings of the Academy of Sciences of the USA* **104** (2007), 8685–8690.

[GDM02] A. V. Goltsev, S. N. Dorogovtsev, and J. F. F. Mendes, Critical phenomena in networks, *Physical Review E* **67** (2002), 026123.

[GDM06] A. V. Goltsev, S. N. Dorogovtsev, and J. F. F. Mendes, k-core (boot-strap) percolation on complex networks: critical phenomena and nonlocal effects, *Physical Review E* **73** (2006), 056101.

[GDM08] A. V. Goltsev, S. N. Dorogovtsev, and J. F. F. Mendes, Percolation on correlated networks, *Physical Review E* **78** (2008), 051105.

[GGA+02] X. Guardiola, R. Guimera, A. Arenas, A. Diaz-Guilera, D. Streib, and L. A. N. Amaral, Macro- and micro-structure of trust networks, ArXiv Condensed Matter e-prints (2002).

[GHB08] M. C. González, C. A. Hidalgo, and A.-L. Barabási, Understanding individual human mobility patterns, *Nature* **453** (2008), 779–782.

[GI95] J. W. Grossman and P. D. F. Ion, On a portion of the well-known collaboration graph, *Congressus Numerantium* **108** (1995), 129–131.

[GKK01a] K.-I. Goh, B. Kahng, and D. Kim, Spectra and eigenvectors of scale-free networks, *Physical Review E* **64** (2001), no. 5, 051903.

[GKK01b] K.-I. Goh, B. Kahng, and D. Kim, Universal behavior of load distribution in scale-free networks, *Physical Review Letters* **87** (2001), 278701.

[GLA+07] L. K. Gallos, F. Liljeros, P. Argyrakis, A. Bunde, and S. Havlin, Improving immunization strategies, *Physical Review E* **75** (2007), 045104.

[GMS03] C. Gkantsidis, M. Mihail, and A. Saberi, Conductance and congestion in power law graphs, *SIGMETRICS*, 2003, pp. 148–159. New York: ACM.

[GN02] M. Girvan and M. E. J. Newman, Community structure in social and biological networks, *Proceedings of the National Academy of Sciences of the USA* **99** (2002), 7821–7826.

[GOJ+02] K.-I. Goh, E. S. Oh, H. Jeong, B. Kahng, and D. Kim, Classification of scale free networks, *Proceedings of the National Academy of Sciences of the USA* **99** (2002), 12583.

[Gra83] P. Grassberger, On the critical behavior of the general epidemic process and dynamical percolation, *Mathematical Biosciences* **63** (1983), 157.

[GSKK06] K.-I. Goh, G. Salvi, B. Kahng, and D. Kim, Skeleton and fractal scaling in complex networks, *Physical Review Letters* **96** (2006), 018701.

[GSM08] L. K. Gallos, C. Song, and H. A. Makse, Scaling of degree correlations and its influence on diffusion in scale-free networks, *Physical Review Letters* **100** (2008), 248701.

[HA87] H. W. Hethcote and J. W. Van Ark, Epidemiological models for heterogeneous populations: proportionate mixing, parameter estimation and immunization programs, *Mathematical Biosciences* **84** (1987), 85–118.

[Hb87] S. Havlin and D. ben-Avraham, Diffusion in disordered media, *Advances in Physics* **36** (1987), 695.

[HBB+05] S. Havlin, L. A. Braunstein, S. V. Buldyrev, R. Cohen, T. Kalisky, S. Sreenivasan, and H. Eugene Stanley, Optimal path in random networks with disorder: a mini review, *Physica A* **346** (2005), 82–92.

[HBG06] L. Hufnagel, D. Brockmann, and T. Geisel, The scaling laws of human travel, *Nature* **439** (2006), 462–465.

[HBP03] C. Herrmann, M. Barthelemy, and P. Provero, Connectivity distribution of spatial networks, *Physical Review E* **68** (2003), 026128.

[HBR96] M. Huxham, S. Beaney, and D. Raffaelli, Do parasites reduce the chances of triangulation in a real food web?, *Oikos* **76** (1996), 284–300.

[HCE07] Y. He, Z. J. Chen, and A. C. Evans, Small-world anatomical networks in the human brain revealed by cortical thickness from MRI, *Cerebral Cortex* **17** (2007), 2407–2419.

[Hof76] D. R. Hofstadter, Energy levels and wave functions of bloch electrons in rational and irrational magnetic fields, *Physical Review B* **14** (1976), 2239–2249.

[Hol04] P. Holme, Efficient local strategies for vaccination and network attack, *Europhysics Letters* **68** (2004), 908–914.

[HSF+05] J. A. Holyst, J. Sienkiewicz, A. Fronczak, P. Fronczak, and K. Suchecki, Scaling of distances in correlated complex networks, *Physica A* **351** (2005), 167–174.

[IA05] S. Itzkovitz and U. Alon, Subgraphs and network motifs in geometric networks, *Physical Review E* **71** (2005), 026117.

[IMK+03] S. Itzkovitz, R. Milo, N. Kashtan, G. Ziv, and U. Alon, Subgraphs in random networks, *Physical Review E* **68** (2003), 026127.

[JKBH08] L. Jahnke, J. W. Kantelhardt, R. Berkovits, and S. Havlin, Wave localization in complex networks with high clustering, *Physical Review Letters* **101** (2008), 175702.

[JKK02] S. Jung, S. Kim, and B. Kahng, Geometric fractal growth model for scale-free networks, *Physical Review E* **65** (2002), 056101.

[JKYR06] L. Jing, H. Keqing, M. Yutao, and P. Rong, Scale free in software metrics, *Proceedings of the 30th Annual International Computer Software and Applications Conference (COMPSAC'06)*, 2006. New York: IEEE.

[JMBO01] H. Jeong, S. Mason, A.-L. Barabási, and Z. N. Oltvai, Lethality and centrality in protein networks, *Nature* **411** (2001), 41.

[JNB03] H. Jeong, Z. Néda, and A. L. Barabási, Measuring preferential attachment in evolving networks, *Europhysics Letters* **61** (2003), 567.

[JTA+00] H. Jeong, B. Tombor, R. Albert, Z. N. Oltvai, and A.-L. Barabási, The large-scale organization of metabolic networks, *Nature* **407** (2000), 651.

[KBB+05] T. Kalisky, L. A. Braunstein, S. V. Buldyrev, S. Havlin, and H. E. Stanley, Scaling of optimal-path-lengths distribution in complex networks, *Physical Review E* **72** (2005), no. 2, 025102.

[KCM+06] T. Kalisky, R. Cohen, O. Mokryn, D. Dolev, Y. Shavitt, and S. Havlin, Tomography of scale-free networks and shortest path trees, *Physical Review E* **74** (2006), 066108.

[KFY04] D. Krioukov, K. Fall, and X. Yang, Compact routing on internet-like graphs, *Proceedings of IEEE INFOCOM*, 2004. New York: IEEE.

[KHB08] K. Kosmidis, S. Havlin, and A. Bunde, Structural properties of spatially embedded networks, *Europhysics Letters* **82** (2008), 48005.

[Kim04] B. J. Kim, Geographical coarse graining of complex networks, *Physical Review Letters* **93** (2004), 168701.

[KIMA04] N. Kashtan, S. Itzkovitz, R. Milo, and U. Alon, Topological generalizations of network motifs, *Physical Review E* **70** (2004), 031909.

[KKK02] L. Kullmann, J. Kertész, and K. Kaski, Time-dependent cross-correlations between different stock returns: a directed network of influence, *Physical Review E* **66** (2002), 026125.

[Kle00a] J. M. Kleinberg, Navigation in a small world, *Nature* **406** (2000), 845.

[Kle00b] J. M. Kleinberg, The small-world phenomenon: an algorithm perspective, *Proceedings of the 32nd ACM Symposium on Theory of Computing*, 2000, pp. 163–170. New York: ACM.

[Knu93] D. E. Knuth, *The Stanford Graphbase: a Platform for Combinatorial Computing*, 1993. New York: ACM.

[KOS+07] J. M. Kumpula, J. P. Onnela, J. Saramäki, K. Kaski, and J. Kertész, Emergence of communities in weighted networks, *Physical Review Letters* **99** (2007), 228701.

[KR01] P. L. Krapivsky and S. Redner, Organization of growing random networks, *Physical Review E* **63** (2001), 066123.

[KR05] P. L. Krapivsky and S. Redner, Network growth by copying, *Physical Review E* **71** (2005), 036118.

[KR07] E. Kenah and J. M. Robins, Second look at the spread of epidemics on networks, *Physical Review E* **76** (2007), 036113.

[KRL00] P. L. Krapivsky, S. Redner, and F. Leyvraz, Connectivity of growing random networks, *Physical Review Letters* **85** (2000), 4629–4632.

[KRW88] J. Koplik, S. Redner, and D. Wilkinson, Transport and dispersion in random networks with percolation disorder, *Physical Review A* **37** (1988), 2619–2636.

[KS87] I. Kanter and H. Sompolinsky, Mean-field theory of spin-glasses with finite coordination number, *Physical Review Letters* **58** (1987), 164–167.

[KSZ96] J. Klafter, M. F. Shlesinger, and G. Zumofen, Beyond Brownian motion, *Physics Today* **49** (1996), 33.

[KWC93] J. O. Kephart, S. R. White, and D. M. Chess, Computers and epidemiology, *IEEE Spectrum* **30** (1993), 26–30.

[KWZZ03] F. Kuhn, R. Wattenhofer, Y. Zhang, and A. Zollinger, Geometric ad-hoc routing: of theory and practice, *Proceedings of the 22nd Annual Symposium on Principles of Distributed Computing*, 2003, pp. 63–72. New York: ACM.

[KZ03] F. Kuhn and A. Zollinger, Ad-hoc networks beyond unit disk graphs, *Proceedings of the 2003 Joint Workshop on Foundations of Mobile Computing, DIALM-POMC*, 2003, pp. 69–78. New York: ACM.

[LBCX03] A. Lakhina, J. Byers, M. Crovella, and P. Xie, Sampling biases in ip topology measurements, *Proceedings of IEEE INFOCOM'03*, 2003. New York: IEEE.

[LCH$^+$06] E. López, S. Carmi, S. Havlin, S. V. Buldyrev, and H. E. Stanley, Anomalous electrical and frictionless flow conductance in complex networks, *Physica D* **224** (2006), 69–76.

[LEA$^+$01] F. Liljeros, C. R. Edling, L. A. N. Amaral, H. E. Stanley, and Y. Åberg, The web of human sexual contacts, *Nature* **411** (2001), 907–908.

[LEA03] F. Liljeros, C. R. Edling, and L. A. N. Amaral, Sexual networks: implications for the transmission of sexually transmitted infections, *Microbes and Infection* **5** (2003), 189–196(8).

[Lév25] P. Lévy, *Calcul des Probabilities*, 1925. Paris: Gauthier Villars.

[LM01] A. L. Lloyd and R. M. May, How viruses spread among computers and people, *Science* **292** (2001), 1316–1317.

[Lov96] L. Lovász, Random Walks on Graphs: A Survey, Combinatorics, Paul Erdős is Eighty (D. Miklś, V. T. Sós, and T. Szőnyi, eds.), volume 2, János Bolyai Mathematical Society, Budapest, 1996, pp. 353–398.

[LTWW94] W. E. Leland, M. S. Taqqu, W. Willinger, and D. V. Wilson, On the self-similar nature of ethernet traffic (extended version), *IEEE/ACM Transactions on Networking* **2** (1994), 1–15.

[LVVZ02] M. Leone, A. Vazquez, A. Vespignani, and R. Zecchina, Ferromagnetic ordering in graphs with arbitrary degree distribution, *European Physical Journal B* **28** (2002), 191–197.

[MA84] R. M. May and R. M. Anderson, Spatial heterogeneity and the design of immunization programs, *Mathematical Biosciences* **72** (1984), 83–111.

[Mag03] D. Magoni, Tearing down the Internet, *IEEE Journal on Selected Areas in Communications* **21** (2003), 949–960.

[Man82] B. B. Mandelbrot, *The Fractal Geometry of Nature*, 1982. San Francisco, CA: Freeman.

[Mar91] N. D. Martinez, Artifacts or attributes? Effects of resolution on the Little Rock Lake food web, *Ecological Monographs* **61** (1991), 367–392.

[MIK$^+$04] R. Milo, S. Itzkovitz, N. Kashtan, R. Levitt, and U. Alon, Response to Comment on "Network motifs: simple building blocks of complex networks" and "Superfamilies of evolved and designed networks," *Science* **305** (2004), 1107d.

[Mil67] S. Milgram, The small world problem, *Psychology Today* **2** (1967), 60–67.

[Mil07] J. C. Miller, Epidemic size and probability in populations with heterogeneous infectivity and susceptibility, *Physical Review E* **76** (2007), 010101.

[MKC$^+$04] N. Madar, T. Kalisky, R. Cohen, D. ben-Avraham, and S. Havlin, Immunization and epidemic dynamics in complex networks, *European Physical Journal B* **38** (2004), 269–276.

[ML02] A. E. Motter and Y.-C. Lai, Cascade-based attacks on complex networks, *Physical Review E* **66** (2002), 065102.

[Mot04] A. E. Motter, Cascade control and defense in complex networks, *Physical Review Letters* **93** (2004), 098701.

[MPV02] Y. Moreno, R. Pastor-Satorras, and A. Vespignani, Epidemic outbreaks in complex heterogeneous networks, *European Physical Journal B* **26** (2002), 521–529.

[MR95] M. Molloy and B. Reed, A critical point for random graphs with a given degree sequence, *Random Structures and Algorithms* **6** (1995), 161.

[MR98] M. Molloy and B. Reed, The size of the giant component of a random graph with a given degree sequence, *Combinatorics, Probability and Computing* **7** (1998), 295.

[MRR$^+$53] N. Metropolis, A. W. Rosenbluth, M. N. Rosenbluth, A. H. Teller, and E. Teller, Equation of state calculations by fast computing machines, *Journal of Chemical Physics* **21** (1953), 1087–1092.

[MS02] S. S. Manna and P. Sen, Modulated scale-free network in Euclidean space, *Physical Review E* **66** (2002), 066114.

[MSOI$^+$02] R. Milo, S. Shen-Orr, S. Itzkovitz, N. Kashtan, D. Chklovskii, and U. Alon, Network motifs: simple building blocks of complex networks, *Science* **298** (2002), 824–827.

[MST06] M. Mihail, A. Saberi, and P. Tetali, Random walks with lookahead on power law random graphs, *Internet Mathematics* **3** (2006), 147–152.

[MT07] A. E. Motter and Z. Toroczkai, Optimization in networks, *Chaos* **17** (2007), 026101.

[MVB$^+$08] D. Mancardi, G. Varetto, E. Bucci, F. Maniero, and C. Guiot, Fractal parameters and vascular networks: facts and artifacts, *Theoretical Biology and Medical Modelling* **5** (2008), no. 1, 12.

[New01a] M. E. J. Newman, Scientific collaboration networks. i. network construction and fundamental results, *Physical Review E* **64** (2001), 016131.

[New01b] M. E. J. Newman, The structure of scientific collaboration networks, *Proceedings of the National Academy of Sciences of the USA* **98** (2001), no. 2, 404–409.

[New02a] M. E. J. Newman, Assortative mixing in networks, *Physical Review Letters* **89** (2002), 208701.

[New02b] M. E. J. Newman, Structure and function of complex networks, *SIAM Review* **45** (2002), 167–256.

[New02c] M. E. J. Newman, The spread of epidemic disease on networks, *Physical Review E* **66** (2002), 016128.

[New03a] M. E. J. Newman, Ego-centered networks and the ripple effect, *Social Networks* **25** (2003), 83–95.

[New03b] M. E. J. Newman, Mixing patterns in networks, *Physical Review E* **67** (2003), 026126.

[New06] M. E. J. Newman, Finding community structure in networks using the eigenvectors of matrices, *Physical Review E* **74** (2006), 036104.

[NFB02] M. E. J. Newman, S. Forrest, and J. Balthrop, Email networks and the spread of computer viruses, *Physical Review E* **66** (2002), 035101.

[NG04] M. E. Newman and M. Girvan, Finding and evaluating community structure in networks, *Physical Review E* **69** (2004), 026113.

[NSW01] M. E. J. Newman, S. H. Strogatz, and D. J. Watts, Random graphs with arbitrary degree distributions and their applications, *Physical Review E* **64** (2001), 026118.

[NW99] M. E. J. Newman and D. J. Watts, Scaling and percolation in the small-world network model, *Physical Review E* **60** (1999), 7332.

[NWS02] M. E. J. Newman, D. J. Watts, and S. H. Strogatz, Random graph models of social networks, *Proceedings of the National Academy of Sciences of the USA* **99** (2002), 2566.

[NZ01] M. E. J. Newman and R. M. Ziff, A fast Monte Carlo algorithm for site or bond percolation, *Physical Review E* **64** (2001), 016706.

[OKK04] J.-P. Onnela, K. Kaski, and J. Kertész, Clustering and information in correlation based financial networks, *European Physical Journal B* **38** (2004), 353–362.

[Ons44] L. Onsager, Crystal statistics. i. a two-dimensional model with an order-disorder transition, *Physical Review* **65** (1944), 117–149.

[Pax97] V. Paxson, End-to-end routing behavior in the internet, *IEEE/ACM Transactions on Networking* **5** (1997), 601–615.

[PBMW99] L. Page, S. Brin, R. Motwani, and T. Winograd, The pagerank citation ranking: Bringing order to the web, *Technical Report*, Stanford InfoLab, 1999.

[PBV07] G. Palla, A.-L. Barabási, and T. Vicsek, Quantifying social group evolution, *Nature* **446** (2007), 664–667.

[PCS+07] G. Paul, R. Cohen, S. Sreenivasan, S. Havlin, and H. E. Stanley, Graph partitioning induced phase transitions, *Physical Review Letters* **99** (2007), no. 11, 115701.

[PN03] J. Park and M. E. J. Newman, Origin of degree correlations in the internet and other networks, *Physical Review E* **68** (2003), 026112.

[PR04] T. Petermann and P. De Los Rios, Exploration of scale-free networks, *European Physical Journal B* **38** (2004), 201.

[PRK03] A. Pikovsky, M. Rosenblum, and J. Kurths, *Synchronization: A Universal Concept in Nonlinear Sciences, Cambridge Nonlinear Science Series*, 2003. Cambridge: Cambridge University Press.

[PSHB99] M. Porto, N. Schwartz, S. Havlin, and A. Bunde, Optimal paths in disordered media: scaling of the crossover from self-similar to self-affine behavior, *Physical Review E* **60** (1999), R2448–R2451.

[PSS05] G. Paul, S. Sreenivasan, and H. E. Stanley, Resilience of complex networks to random breakdown, *Physical Review E* **72** (2005), 056130.

[PTHS04] G. Paul, T. Tanizawa, S. Havlin, and H. E. Stanley, Optimization of robustness of complex networks, *European Physical Journal B* **38** (2004), 187–191.

[PU89] D. Peleg and E. Upfal, A tradeoff between space and efficiency for routing tables, *Journal of the ACM* **36** (1989), 510–530.

[PV01a] R. Pastor-Satorras and A. Vespignani, Epidemic dynamics and endemic states in complex networks, *Physical Review E* **63** (2001), 066117.

[PV01b] R. Pastor-Satorras and A. Vespignani, Epidemic spreading in scale-free networks, *Physical Review Letters* **86** (2001), 3200–3203.

[PV02] R. Pastor-Satorras and A. Vespignani, Immunization of complex networks, *Physical Review E* **65** (2002), 036104.

[PV03] R. Pastor-Satorras and A. Vespignani, *Evolution and Structure of the Internet: a Statistical Physics Approach*, 2003. Cambridge: Cambridge University Press.

[PVV01] R. Pastor-Satorras, A. Vázquez, and A. Vespignani, Dynamical and correlation properties of the internet, *Physical Review Letters* **87** (2001), 258701.

[PW00] K. Park and W. Willinger, *Self-Similar Network Traffic and Performance Evaluation*, 2000. New York: Wiley.

[Rap57] A. Rapoport, A contribution to the theory of random and biased nets, *Bulletin of Mathematical Biophysics* **19** (1957), 257–271.

[RB03] E. Ravasz and A.-L. Barabási, Hierarchical organization in complex networks, *Physical Review E* **67** (2003), 026112-+.

[Rb04] H. D. Rozenfeld and D. ben-Avraham, Designer nets from local strategies, *Physical Review E* **70** (2004), 056107.

[Rb06] M. R. Roberson and D. ben-Avraham, Kleinberg navigation in fractal small-world networks, *Physical Review E* **74** (2006), 017101.

[RCbH02] A. F. Rozenfeld, R. Cohen, D. ben-Avraham, and S. Havlin, Scale free networks on lattices, *Physical Review Letters* **89** (2002), 218701.

[Red98] S. Redner, How popular is your paper? An empirical study of the citation distribution, *European Physical Journal B* **4** (1998), 131–134.

[RFI02] M. Ripeanu, I. Foster, and A. Iamnitchi, Mapping the Gnutella network: properties of large-scale peer-to-peer systems and implications for system design, *IEEE Internet Computing* **6** (2002), 50–57.

[RGMS05] M. Rosvall, A. Gronlund, P. Minnhagen, and K. Sneppen, Searchability of networks, *Physical Review E* **72** (2005), 046117.

[RHb07] H. D. Rozenfeld, S. Havlin, and D. ben-Avraham, Fractal and transfractal recursive scale-free nets, *New Journal of Physics* **9** (2007), 175.

[RMS05] M. Rosvall, P. Minnhagen, and K. Sneppen, Navigating networks with limited information, *Physical Review E* **71** (2005), 066111.

[RSM+02] E. Ravasz, A. L. Somera, D. A. Mongru, Z. N. Oltvai, and A.-L. Barabási, Hierarchical organization of modularity in metabolic networks, *Science* **297** (2002), 1551–1555.

[SA94] D. Stauffer and A. Aharony, *Introduction to Percolation Theory*, second edition, 1994. London: Taylor and Francis.

[SAK02] G. Szabó, M. Alava, and J. Kertész, Shortest paths and load scaling in scale free trees, *Physical Review E* **66** (2002), 026101.

[SBC+08] J. Shao, S. V. Buldyrev, R. Cohen, M. Kitsak, S. Havlin, and H. E. Stanley, Fractal boundaries of complex networks, *Europhysics Letters* **84** (2008), 48004.

[SC01] P. Sen and B. K. Chakrabarti, Small-world phenomena and the statistics of linear polymers, *Journal of Physics A* **34** (2001), 7749–7755.

[SCbA+02] N. Schwartz, R. Cohen, D. ben Avraham, S. Havlin, and A.-L. Barabási, Percolation in directed scale-free networks, *Physical Review E* **66** (2002), 015104(R).

[SCL+07] S. Sreenivasan, R. Cohen, E. Lopez, Z. Toroczkai, and H. E. Stanley, Structural bottlenecks for communication in networks, *Physical Review E* **75** (2007), 036105.

[SDC+03] P. Sen, S. Dasgupta, A. Chatterjee, P. A. Sreeram, G. Mukherjee, and S. S. Manna, Small-world properties of the Indian railway network, *Physical Review E* **67** (2003), 036106.

[SDM08] A. N. Samukhin, S. N. Dorogovtsev, and J. F. F. Mendes, Laplacian spectra of, and random walks on, complex networks: are scale-free architectures really important?, *Physical Review E* **77** (2008), 036115.

[SGHM07] C. Song, L. K. Gallos, S. Havlin, and H. A. Makse, How to calculate the fractal dimension of a complex network: the box covering algorithm, *Journal of Statistical Mechanics: Theory and Experiment* **2007** (2007), P03006.

[SHBF05] Y. M. Strelniker, S. Havlin, R. Berkovits, and A. Frydman, Resistance distribution in the hopping percolation model, *Physical Review E* **72** (2005), 016121–+.

[SHM05] C. Song, S. Havlin, and H. A. Makse, Self-similarity of complex networks, *Nature* **433** (2005), 392–395.

[SHM06] C. Song, S. Havlin, and H. A. Makse, Origins of fractality in the growth of complex networks, *Nature Physics* **2** (2006), 275–281.

[Sin93] A. Sinclair, *Algorithms for Random Generation and Counting: A Markov Chain Approach, progress in theoretical computer science*, 1993. Boston, MA: Birkhäuser.

[SK85] M. F. Shlesinger and J. Klafter, Accelerated diffusion in Josephson junctions and related chaotic systems, *Physical Review Letters* **54** (1985), 2551.

[SKHB05] M. Sade, T. Kalisky, S. Havlin, and R. Berkovits, Localization transition on complex networks via spectral statistics, *Physical Review E* **72** (2005), 066123.

[SKO+07] J. Saramaki, M. Kivela, J.-P. Onnela, K. Kaski, and J. Kertesz, Generalizations of the clustering coefficient to weighted complex networks, *Physical Review E* **75** (2007), 027105.

[SM01] R. V. Sole and J. M. Montoya, Complexity and fragility in ecological networks, *Proceedings of the Royal Society of London, Series B* **268** (2001), 2039.

[SOMMA02] S. Shen-Orr, R. Milo, S. Mangan, and U. Alon, Network motifs in the transcriptional regulation network of escherichia coli, *Nature Genetics* **31** (2002), 64–68.

[SS96] M. Sipser and D. A. Spielman, Expander codes, *IEEE Transactions on Information Theory* **42** (1996), 1710–1722.

[SS05] Y. Shavitt and E. Shir, Dimes: let the internet measure itself, *Computer Communication Review* **35** (2005), 71–74.

[Str00] S. H. Strogatz, Exploring complex networks, *Nature* **410** (2000), 268–276.

[Str03] S. Strogatz, *Sync: The Emerging Science of Spontaneous Order*, 2003. New York: Hyperion.

[STR05] K. Sneppen, A. Trusina, and M. Rosvall, Hide-and-seek on complex networks, *Europhysics Letters* **69** (2005), 853–859.

[SZK93] M. F. Shlesinger, G. M. Zaslavsky, and J. Klafter, Strange kinetics, *Nature* **363** (1993), 31–37.

[TPC+05] T. Tanizawa, G. Paul, R. Cohen, S. Havlin, and H. E. Stanley, Optimization of network robustness to waves of targeted and random attacks, *Physical Review E* **71** (2005), 047101.

[TRS05] A. Trusina, M. Rosvall, and K. Sneppen, Communication boundaries in networks, *Physical Review Letters* **94** (2005), 238701.

[TZ01] M. Thorup and U. Zwick, Compact routing schemes, *Proceedings of the 13th Annual ACM Symposium on Parallel Algorithms and Architectures*, 2001. New York: ACM.

[Vaz04] V. V. Vazirani, *Approximation Algorithms*, 2004. Berlin: Springer.

[VBB+02] G. M. Viswanathan, F. Bartumeus, S. V. Buldyrev, J. Catalan, U. L. Fulco, S. Havlin, M. G. E. da Luz, M. L. Lyra, E. P. Raposo, and H. E. Stanley, Levy flight random searches in biological phenomena, *Physica A* **314** (2002), 208–213.

[VBH+99] G. M. Viswanathan, S. V. Buldyrev, S. Havlin, M. G. da Luz, E. P. Raposo, and H. E. Stanley, Optimizing the success of random searches, *Nature* **401** (1999), 911–914.

[VCS02] S. Valverde, R. Ferrer Cancho, and R. V. Solé, Mixing patterns in networks, *Europhysics Letters* **60** (2002), 512–517.

[vdHHvM01] R. van der Hofstad, G. Hooghiemstra, and P. van Mieghem, First passage percolation on the random graph, *Probability in the Engineering and Informational Sciences* **15** (2001), 225–237.

[VPV02] A. Vazquez, R. Pastor-Satorras, and A. Vespignani, Internet topology at the router and autonomous system level, *Condensed Matter* (2002), 0206084.

[VSS04] A. X. C. N. Valente, A. Sarkar, and H. A. Stone, Two-peak and three-peak optimal complex networks, *Physical Review Letters* **92** (2004), 118702.

[Wat99] D. J. Watts, *Small Worlds*, 1999. Princeton, NJ: Princeton University Press.

[Wat03] D. J. Watts, *Six Degrees: The Science of a Connected Age*, 2003. New York: W. W. Norton.

[WBC+06] Z. Wu, L. A. Braunstein, V. Colizza, R. Cohen, S. Havlin, and H. E. Stanley, Optimal paths in complex networks with correlated weights: the worldwide airport network, *Physical Review E* **74** (2006), 056104.

[WBE97] G. B. West, J. H. Brown, and B. J. Enquist, A general model for the origin of allometric scaling laws in biology, *Science* **276** (1997), 122–126.

[WBHS06] Z. Wu, L. A. Braunstein, S. Havlin, and H. E. Stanley, Transport in weighted networks: partition into superhighways and roads, *Physical Review Letters* **96** (2006), 148702.

[WDN02] D. J. Watts, P. S. Dodds, and M. E. J. Newman, Identity and search in social networks, *Science* **296** (2002), 1302–1305.

[Wei94] G. H. Weiss, *Aspects and applications of the random walk*, 1994. Amsterdam: North-Holland.

[WGHB09] P. Wang, M. C. Gonzalez, C. A. Hidalgo, and A.-L. Barabási, Understanding the spreading patterns of mobile phone viruses, *Science* **324** (2009), 1071–1076.

[Wig55] E. P. Wigner, Characteristic vectors of bordered matrices with infinite dimensions, *Annals of Mathematics* **62** (1955), 548–564.

[Wil94] H. S. Wilf, *Generatingfunctionology*, 1994. London: Academic Press.

[WS98] D. J. Watts and S. H. Strogatz, Collective dynamics of "small world" networks, *Nature* **393** (1998), 440–442.

[WSS02a] C. P. Warren, L. M. Sander, and I. M. Sokolov, Firewalls, disorder, and percolation in networks, *Mathematical Biosciences* **180** (2002), 293.

[WSS02b] C. P. Warren, L. M. Sander, and I. M. Sokolov, Geography in a scale-free network model, *Physical Review E* **66** (2002), 56105.

[WSTB86] J. G. White, E. Southgate, J. N. Thomson, and S. Brenner, The structure of the nervous system of the nematode *caenorhabditis elegans*, *Philosophical Transactions of the Royal Society of London, Series B* **314** (1986), 1–340.

[WY84] W. H. Wethcote and J. A. Yorke, *Gonorrhea Transmission Dynamics and Control*, Lecture notes in Biomathematics, volume 56, 1984. Berlin: Springer.

[XSD+02] I. Xenarios, L. Salwínski, X. J. Duan, P. Higney, S.-M. Kim, and D. Eisenberg, DIP, the database of interacting proteins: a research tool for studying cellular networks of protein interactions, *Nucleic Acids Research* **30** (2002), 303–305.

[YJB02] S.-H. Yook, H. Jeong, and A.-L. Barabási, Modeling the Internet's large-scale topology, *Proceedings of the National Academy of Sciences of the USA* **99** (2002), 13382–13386.

[ZKM+03] A. Zilman, J. Kieffer, F. Molino, G. Porte, and S. A. Safran, Entropic phase separation in polymer-microemulsion networks, *Physical Review Letters* **91** (2003), 015901.

Index